Adapt or Die

Adapt or Die

The science, politics and economics of climate change

Edited by

Kendra Okonski

P

PROFILE BOOKS

First published in Great Britain in 2003 by
Profile Books Ltd
58A Hatton Garden
London ECIN 8LX
www.profilebooks.co.uk

in association with
International Policy Network

10 9 8 7 6 5 4 3 2 1

The foreword, introduction, essays and summary in this collection are published here for the first time and are copyright of the author in each case.

Designed and typeset in Sabon by MacGuru
info@macguru.org.uk

Printed and bound in Great Britain by
Clays, Bungay, Suffolk

The moral right of the authors has been asserted.

A CIP catalogue record for this book is available from the British Library.

ISBN 1 86197 795 6

5000069251

U H University of Hertfordshire

College Lane, Hatfield, Herts. AL10 9AB

Learning and Information Services

For renewal of Standard and One Week Loans,
please visit the web site **http://www.voyager.herts.ac.uk**

This item must be returned or the loan renewed by the due date.
The University reserves the right to recall items from loan at any time.
A fine will be charged for the late return of items.

Contents

The authors

Martin Ågerup is a Bachelor of Science in economics and economic history from the University of Bristol and MA in European economics from the University of Exeter. He has written a best-selling Danish book about environment issues. Martin is President of the Danish Academy for Futures Studies.

Stuart Buck is an appellate attorney with Hughes & Luce in Dallas, Texas. He holds a law degree from Harvard Law School, where he was an editor of the *Harvard Law Review*. He subsequently clerked for Judge David A. Nelson of the US Court of Appeals (Sixth Circuit) and Judge Stephen F. Williams (US Court of Appeals, DC Circuit). His scholarly articles have appeared in the *Harvard Law Review*, the *Stanford Technology Law Review*, the *Harvard Environmental Law Review*, the *Utah Law Review* and others.

Indur M. Goklany is an engineer and a visiting fellow at the American Enterprise Institute. He is the author of numerous books and papers on climate change, biodiversity and the relationship between economic growth and the environment.

Andrew Kenny is a professional engineer from South Africa with degrees in physics and mechanical engineering. He has worked in industry in England and South Africa for seventeen years, including periods spent at a coal power station and the nuclear engineering department of Eskom, the South African electricity utility. For the last five years he has been working in energy research at the University of Cape Town's

Energy Research Institute. He is also a freelance journalist specialising in political and environmental themes.

Martin Livermore is a chemist by training, has a background in the food and biotechnology industries and now works as an independent consultant. He has key interests in a range of environmental and sustainability issues, including climate change, and works with a number of organisations to promote scientific rationality in policy making and risk management. Contact address: martinlivermore@aol.com

Barun S. Mitra is a founder and Executive Director of Liberty Institute in New Delhi, India, a non-governmental, non-profit organisation. Mitra is a frequent writer and commentator on issues relating to development, trade, environment and technology. His writings have been published extensively and he has contributed to several books.

Julian Morris is Executive Director of International Policy Network (London) and a visiting professor at the University of Buckingham. An economist and a lawyer, he has written or edited several books and academic papers.

Kendra Okonski is Director of Sustainable Development for International Policy Network (London). She has written widely for the international media, and holds a degree in economics from Hillsdale College. Contact: kokonski@policynetwork.net.

Dr Benny Peiser is a social anthropologist at Liverpool John Moores University (UK). His research interest focuses on the effects of catastrophic events on societal evolution and the emergence of apocalyptic movements and religions. Benny is a fellow of the Royal Astronomical Society, an advisor to the Scientific Alliance, and the editor of The Cambridge Conference Network (CCNet).

Dr Paul Reiter is a specialist in mosquitoes, the diseases they transmit and improved methods for mosquito control. He worked for 22 years as a medical entomologist for the Division of Vector-Borne Infectious Disease of the Centers for Disease Control and Prevention (CDC) and now heads a new unit of Insects and Infectious Disease at the Pasteur Institute in Paris.

Dominic Standish is completing a sociological PhD at the University of Kent (UK) on environmental risks in Venice, and teaches and writes on related issues, including work for a Consortium of American Universities (CIMBA) based in Asolo, Italy. He is also a freelance journalist for publications such as the *International Herald Tribune* and several online websites. He lives near Venice with his wife and two sons. Contact: dstandish@europe.com

Carlo Stagnaro, a freelance journalist based in Italy, is a Fellow of International Policy Network (London) and co-edits the magazine *Enclave* (Treviglio). His articles on environmental issues have appeared in several newspapers and magazines across Italy. He is also a contributor for the specialised press agency GreenWatchNews (Rome) and director of the series 'Libera scienza' for the Leonardo Facco Editore publishing firm. He is a member of the society 'Galileo 2001 – For the Freedom and Dignity of Science'.

Dr Bruce Yandle is Professor of Economics Emeritus, Clemson University, Clemson, SC, USA, and senior associate with the Political Economy Research Center (PERC), Bozeman, Montana, USA. A member of the editorial board of the *European Journal of Law & Economics*, he is author/editor of a dozen books on environmental policy. Yandle served as Executive Director of the US Federal Trade Commission, 1982–84.

Tables and figures

Tables

Figures

Abbreviations

CDM	Clean Development Mechanism
CO_2	carbon dioxide
COP	Conference of Parties
DEFRA	UK Department of Environment, Food and Rural Affairs
DFID	UK Department for International Development
DOE	US Department of Energy
EC	European Commission
EEA	European Environment Agency
EPA	US Environmental Protection Agency
EU	European Union
FAO	Food and Agriculture Organization
GATT	General Agreement on Tariffs and Trade
GDP	Gross Domestic Product
GHG	Greenhouse gases
IEA	International Energy Agency
IPCC	Intergovernmental Panel on Climate Change
LPG	Liquid Petroleum Gas
MEA	Multilateral Environmental Agreement
MOSE	Modulo Sperimentale Elettromeccanico (Experimental Electromechanical Module)
NGO	Non-governmental organisation
OECD	Organization for Economic Cooperation and Development
SMEs	Small and Medium-sized Enterprises
SRES	Special Report on Emissions Scenarios
TAR	Third Assessment Report of the Intergovernmental Panel on Climate Change
UNCED	1992 United Nations Conference on Environment and Development, Rio de Janeiro

UNDP	United Nations Development Programme
UNEP	United Nations Environment Programme
UNFCCC	United Nations Framework Convention on Climate Change
WHO	World Health Organization
WSSD	2002 World Summit on Sustainable Development, Johannesburg
WTO	World Trade Organization

The terms 'climate change' and 'global warming' are used interchangeably in this book.

'Kyoto Protocol' and 'Kyoto' are also used interchangeably.

Foreword

This book is both important and timely because it confronts one of the most tenacious myths of the new millennium, namely the dangerously mistaken belief that the correct approach to climate change is to try to 'manage' climate itself. The authors of this volume do not challenge the idea that climate changes, and has always changed. Nor do they dispute the fact that humans may have some role in this change, however complex and difficult the human impact might be to unravel scientifically.

What they are saying is of far more practical and immediate significance. First, they remind us that the age-old response to climate change has not been the hubris of 'control', but rather a constant Darwinian adaptation to change, whatever its direction – warm, wet, cold, dry, or all at once. Secondly, they argue cogently that the current responses to climate change adopted by the Intergovernmental Panel on Climate Change (IPCC), the European Union (EU), and certain non-governmental organisations and businesses, far from aiding the process of political, social and economic adaptation, may already be starting to undermine our future ability to remain adaptive. As Martin Livermore puts it with respect to the EU: 'Present EU energy policy is leading us into possibly the worst of all possible worlds: a strategy which is unnecessary, because the science of climate change is uncertain; ineffective, because the EU's cuts in emissions will not stave off a warmer climate; and negative for nearly everyone in society.'

And the negative outcomes for society could be apparent far sooner than we might expect. In the UK, for example, the Institution of Civil Engineers (ICE) has recently issued a stark warning that the government's energy policy, predicated on cloud-cuckoo-land action to stop 'global warming', is likely to lead to severe power cuts, and homes without heat and light, by 2020. Some industry experts have even argued that this could happen as early as the winter of 2005–6. And how will a weakened, energy-poor economy cope then with change? Bruce Yandle and Stuart Buck tellingly point out that even 'if global warming turns out to be genuine, those economies that maintain market flexibility will be best equipped to adapt to it'.

This is why it is so vital in the present debate to distinguish sharply between the grand narrative of 'global warming' and the demanding reality of climate change. The myth of 'global warming' has beguiled far too many of our well-meaning scientists and policy makers into a false hope that we can create a 'sustainable' climate. The very concept of a 'sustainable' climate is an oxymoron. It is a foolishness of quixotic proportions to think that we can manage climate in any *predictable* manner by trying to fiddle at the margins with just a couple of factors out of the millions that interact to determine the direction of climate change. By accident, we could even plunge ourselves into another ice age.

Wisely, the writers in this volume sweep aside such pie-in-the-sky fantasies and look instead at down-to-earth ways in which we can and must continue to adapt to inexorable climate change if we are to survive. In doing so, the authors also dismiss a whole suite of secondary myths that have fed off the main 'global warming' narrative. Nowhere is this more so than in the first section of the book, which analyses the possible impacts of climate change. Paul Reiter brilliantly debunks the nonsense that present-day climate change will automatically increase the incidence of tropical diseases, while Dominic Standish unravels the whole sorry tale of Venice and the climate change debate in Italy. By contrast, Indur Goklany examines positive ways in which we may enhance our ability to adapt to climate change, through technology, economic growth and free trade.

In the second part, Martin Ågerup, Barun S. Mitra and Andrew Kenny lay bare the deception that is the Kyoto Protocol. Of particular note are the analyses of the potential neo-colonial impacts of the Protocol on the developing world, where at least 1.6 billion people remain starved of energy. As Mitra concludes, 'Reliance on traditional energy has many disadvantages and is responsible for enormous human suffering, loss of life, reduced economic productivity and environmental problems.' But, as both Mitra and Kenny illustrate, the Clean Development Mechanism – part of the Kyoto Protocol – 'because it is motivated by the wrong goals' is 'likely to hinder rather than promote sustainable development'.

The book then goes on to examine the impact of the 'global warming' mentality on trade (Julian Morris) and on European risk regulation in business (Martin Livermore). Lastly, three authors place present-day climate change policy in a wider set of contexts. Bruce Yandle and Stuart Buck are excoriating in their exposure of the vested interests in 'global warming' exhibited by certain governments, big corporations, and environmental pressure groups, while Benny Peiser locates the whole debate in the context of past civilisations and climate change. In conclusion, Carlo Stagnaro draws all together using the principles of political economy to expose the interest groups that are driving climate change policy in Europe, that part of the world most bewitched and befuddled by the 'global warming' myth.

Kendra Okonski is to be warmly congratulated on bringing together such a telling critique of the science, the politics and the economics of climate change. Throughout the world, the myth of 'global warming' is at last being challenged by those who understand that the only effective way to approach climate change is to maintain strong economies that are flexible and adaptive, not to try to hobble economic growth and energy development for ineffective ends.

Most interestingly, the debate is particularly fierce in Russia, where the leading newspaper, *Pravda*, recently published a strongly worded article entitled 'The Kyoto Protocol is not worth a thing'. Personally, I think it is worse than this: I believe the Protocol is dangerous for humanity because it undermines our fundamental need to grow and to adapt. In the UK, the false agenda is already leading to energy policies that could be disastrous for both people and the environment.

Thus, following the ancient example of King Canute, we must urgently acknowledge that predictable climate control remains beyond our reach; as ever, we must adapt or die. Let us be foxes, not giant pandas.

Philip Stott
Professor Emeritus of Biogeography in the
University of London, Kent, UK
July 2003

Introduction

Kendra Okonski

Science, or progress in science, may be regarded as a means used
by the human species to adapt itself to the environment.

Karl Popper, *The Rationality of Scientific Revolutions*, 1975

Adapt or perish, now as ever, is nature's inexorable imperative.

H. G. Wells

Is this just another book about global warming?

Adapt or Die proposes a framework for thinking about how to respond to potential threats to humanity and the earth which might result from change. Climate change, and in particular global warming,
is the subject of this book, but the proposed framework could apply
equally well to other global or local environmental problems.

Human welfare is inextricably linked to the earth's climate, and
just like all other life forms, the manner in which we respond to
change is critical not only to our survival, but to our well-being. The
prospect of global warming offers an opportunity to re-evaluate how
we address global problems, and how we ought to prioritise our efforts to cope with them.

The debate about climate change has been largely driven by the assumption that the impacts of global warming, if unmitigated, will
pose threats to humanity and to the environment. These threats could
include rising sea levels, changes in agricultural production, severe
weather events, the spread of disease, and environmental effects such
as the loss of biodiversity.

Those who are involved in formulating climate policy at national
and international levels are faced with the enormous task of evaluating

the evidence, and devising possible solutions to global warming. Most have taken the view that humanity's emissions of greenhouse gases have exacerbated the earth's natural greenhouse effect, which has in turn led to warmer temperatures. Thus, the conventional wisdom suggests that temperatures will continue to rise so long as we continue to emit greenhouse gases and the only way to deal with threats is to drastically reduce our emissions.

At the international level, the Intergovernmental Panel on Climate Change (IPCC) was established to evaluate the extent of the problem, and proposed solutions such as the Kyoto Protocol, an agreement negotiated during the 1990s as the primary global strategy to address global warming. The IPCC has produced three assessment reports, and as of writing the Kyoto Protocol has not yet gone into force worldwide.

Kyoto has often been called an 'insurance policy' against global warming. But against what exactly is it insuring, and at what price? It is not clear that Kyoto will have any effect in terms of reducing global temperatures – and thus would it actually be an effective insurance policy? A discussion of adaptation measures, and the institutional framework to support adaptation, has been largely missing from this debate. More broadly, and as this book's authors show, it is uncertain that mitigation strategies are the best way to reduce our vulnerability to the impacts of a warmer climate.

About this book

The spirit of this book is implied by its title: *Adapt or Die*. Humanity has, over the past 12,000 years (since the advent of fire, the wheel and modern agriculture), adapted to changing circumstances, and there is no reason to believe that such adaptation will not continue. The challenge is to create an environment in which adaptation can flourish without being stifled.

The book's contributors evaluate proposed strategies to mitigate global warming and its potential consequences and compare these with strategies focused on adaptation. In various ways, they suggest that we ought to prioritise our efforts and responses – mainly by reducing the vulnerability of humanity to the effects of a warmer climate in a way which promotes maximum flexibility to cope with change, and with the fewest negative consequences for human well-being.

What this book is not

This book is not an attempt to disprove that global warming is occurring. While there is good reason to have an informed, rational debate about global warming, individual contributors may be more or less convinced that global warming is occurring, or that models of climatic and economic change are accurate in their predictions. Those authors have expressed their own views based on evidence which they believe is important and relevant. All the contributors share a belief that policy makers, the news media and the public should have a better understanding of the evidence for and the potential effects of a warmer climate.

Likewise, the contributors to this book do not suggest that we simply do nothing to respond to climate change. They believe that we should carefully consider our approaches to potential problems which could result from a warmer climate.

The book does not credit, blame or implicate any particular interest group – whether business, government, environmental groups, scientists or the news media – for the state of climate debate. However, it is important to understand the role of those groups in the formation of climate policy. Certain vested interests benefit from climate policy, while the costs are widely dispersed to consumers, small businesses and taxpayers.

This book's contributors do not suggest particular outcomes, nor do they advocate particular technologies to deal with a warmer climate. They show that new technologies of all kinds are desirable to deal with change, including global warming, and in various ways they have suggested an institutional framework which would encourage the development of and access to new technologies, but avoid the consequences of outcome-based sustainable development.

Structure of the book

Adapt or Die is divided into several parts which examine various aspects of the climate change debate. Since much of the debate has promoted the potential negative impacts and consequences of global warming, Part I examines those claims, including a specific focus on disease and the historic city of Venice.

In Part II, the authors evaluate various strategies and policies which have been pursued to deal with climate change – both those of 'climate control' and those of adaptation – given knowledge and expectations about the future. Two authors in this section focus on how climate policy could potentially affect poor countries.

Part III examines the economic consequences of climate policy for businesses and international trade.

To conclude the book, Part IV analyses climate policy using political economy principles, and evaluates justifications for climate policy based on how past civilisations may have been affected by a changing climate.

Part One: Possible Impacts of Climate Change

Perhaps the most ominous aspect of global warming is the negative effects that have been promoted as its consequences. The conventional wisdom is that storms will damage our seafront properties, low-lying countries will be inundated, and that malaria and other diseases will spread to Europe and North America. Part One examines these claims in detail.

Indur Goklany (Chapter 3) analyses potential human and environmental vulnerabilities to the impacts of climate change. First, he examines the consequences of climate change in the past few decades, arguing that:

> Despite any warming, by virtually any climate-sensitive measure of
> human well-being, the average person's welfare has improved over
> the last century ... Most of these improvements are due to
> technological progress driven by market- and science-based
> economic growth, technology and trade.

He examines the possible future impacts of a warmer climate, including extreme weather, sea-level rise, biodiversity loss and the spread of disease, and then analyses the relative merits and disadvantages of pursuing strategies based on mitigation and adaptation.

One of the impacts claimed by some is the spectre of the spread of mosquito-borne diseases, such as malaria, yellow fever and dengue, to Europe and North America as a consequence of a warmer climate. Interest groups have utilised such claims to bolster their calls for urgent mitigation strategies to control the earth's climate.

Paul Reiter (Chapter 1) analyses these claims, in the historical context of malaria in Europe. Malaria was prevalent in ancient Greece and Rome, and across Europe until the second half of the nineteenth century, when it began to decline in much of northern Europe. But its decline, says Reiter, ' ... cannot be attributed to climate change, for it occurred during a warming phase, when temperatures were already much higher than in the Little Ice Age'.

Reiter points to numerous factors which contributed to the decline of malaria in northern Europe, such as new farming practices, which helped to eliminate mosquito habitat; new livestock breeding processes, which separated livestock from human beings; improved human living conditions due to better construction methods and improved materials; and better medical care.

Reiter suggests that:

> [while] there is much talk of efforts to improve the health of
> poorer nations, at the same time, erroneous concepts of mosquito-
> borne disease are used as an argument to spend colossal amounts
> of scarce resources to 'halt' global warming, even as climate
> experts confess that the true contribution of human activities to
> the present warming trend is uncertain.

Dominic Standish (Chapter 2) reviews how Venice has become an emotive issue in the climate change debate. Unlike traditional cities, Venice consists of a series of canals in a lagoon which borders the Mediterranean Sea. This fact, combined with its Renaissance architecture and plundered treasures, have made it a place of wonder for many. Sadly, Venice's construction has made it vulnerable to flooding – especially because it appears to be sinking. But now some are claiming that rising sea levels might be making things worse.

Unfortunately, this has created a strange conflict for Venice, because certain groups in Italy oppose physical measures which would help to protect the city from flooding – specifically, a system of mobile barriers in the lagoon called Project MOSE.

Project MOSE is intended to save this historic – but artificial – city. Opponents, including environmental groups and campaigners, say that the project is 'unnatural' and 'risky'. Yet without such measures, Venice may be more drastically affected by rising sea levels, and also subsidence. Standish believes that this strange dilemma has resulted because 'balanced risk assessments have been replaced by the precautionary principle, which encourages extreme caution but may indeed exacerbate safety concerns by attempting to deny that risks exist.'

Part Two: Strategies for Adapting to Climate Change

In Part Two, the authors evaluate various strategies which have been proposed to solve the problem of climate change. Taking into account our present knowledge, and uncertainties about the future, they ask whether we know enough today to understand the potential human welfare impacts of climate change and climate policy. They question whether actions taken today to limit carbon emissions are justified, especially in light of various real problems today which could be solved in part by cleaner energies of all forms.

The most comprehensive policy solution proposed to climate change is the Kyoto Protocol. So-called 'industrialised' countries in Annex I to the agreement have committed to reducing their emissions of greenhouse gases, carbon dioxide in particular. Martin Ågerup (Chapter 4) analyses Kyoto, and asks whether it is 'a good idea', given reasonable expectations of future conditions, based on the best available current knowledge.

He shows that the potential benefits of Kyoto are extremely uncertain, not least because its effects on climate will be negligible. Likewise, the costs of Kyoto will be extremely high, especially for Europeans between 2008–12, when the first Kyoto commitment period ends. He concludes: 'Without any clear benefits, Kyoto could actually be bad for human welfare, even if it had no costs.'

There has been much discussion about policy strategies after the Kyoto Protocol expires in 2012. Ågerup speculates that:

> Given the mechanisms of the Kyoto Protocol and the current political agenda, it's possible that future commitment periods will imply deeper cuts in GHG emissions in order to stabilise CO_2 emissions or global temperatures at some future level by some future date.

However, it is currently impossible to evaluate the costs of a post-Kyoto regime, because it depends entirely on the quantity of emissions cuts, the methods for cutting emissions and the time frame in which such policies are pursued.

Climate and poor countries

The IPCC and others in the climate debate have argued that poor countries will be more vulnerable to the impacts of a warmer climate. Several contributors have a keen interest in ensuring that climate policies pursued by wealthy countries do not adversely affect poor countries.

For the majority of people on the planet, 'sustainable development' means economic development and all the benefits it brings: escaping from subsistence agriculture, less hunger, better health and longer lives, as well as greater creature comforts. This also includes the development of new, more efficient technologies, through voluntary actions of the private sector rather than through government subsidies.

Many people in wealthy countries experience daily the benefits of clean, reliable and affordable energy. They may be unfamiliar with the lifestyles of extremely poor people, and in particular the dire need for cleaner, more efficient and affordable sources of energy in poor countries.

Barun Mitra (Chapter 5) explores how millions of poor people in India rely on low-intensity traditional fuels – mostly wood and cow dung – and how regulations have prevented them from accessing cleaner energy. Mitra argues that:

> There are huge opportunity costs which have not been given
> enough consideration in the climate change debate – particularly
> because these are urgent problems today rather than hypothetical
> long-run problems that might be caused by global warming.

He shows that traditional fuels cause indoor air pollution, which contributes heavily to about 2 million childhood deaths in poor countries every year as well as to poor health for women and young children. Likewise, the use of traditional fuels is contributing to environmental problems such as erosion, deforestation and loss of biodiversity.

Reliance on traditional energy also means that women and children spend much of their time collecting wood or making dung cakes, and thus cannot spend time in other more valuable economic activities or in school.

Mitra argues that poor people need to consume more energy to improve their lifestyle and health, to add value to their economic activities and to eliminate poverty. He believes that

The immediate need of poor people in India and other poor
countries is to consume more energy, in any form ... More energy
consumption will lead to more energy efficiency, which will lead
to environmental benefits and sustainable energy consumption ...
A truly 'clean' path of development would involve little
government, and instead would rely on the initiative and ingenuity
of people to solve energy needs.

Andrew Kenny (Chapter 6) argues that the global-warming debate is
being used to mislead poor countries into pursuing policies that will
harm their development, and thus perpetuate poverty.

Kenny contends that:

Climate change now looms over all official considerations of
energy use at both international and national levels, and it is
casting a shadow over the energy policies of poor countries, who
are being encouraged to buy into agreements which will ultimately
limit their energy consumption.

He says that there is 'excessive anxiety about industrial pollution ...
[but very little concern for] household pollution', in particular pollution which results from fuels (such as paraffin) used for home cooking
and lighting in Africa, Asia and Latin America. The health hazards resulting from these fuels are much larger than those posed by large
power stations or industry.

Kenny also discusses the United Nations' Clean Development
Mechanism (CDM), which is part of the Kyoto Protocol and will
allow rich countries to meet their own emissions cuts through projects
that reduce emissions in poor countries. However, he believes that

The most ominous feature of the [Clean Development Mechanism]
... is the possibility of choosing entirely unsuitable CDM projects
based more on the interests and ideology of the wealthy supplier
than of the poor recipient.

Part Three: The Economic Consequences of Climate Change Policy

This section evaluates the economic consequences of both 'climate
control' policy and alternative strategies to deal with global warming.

In particular, the authors evaluate the intended and unintended effects that climate policy could have on the ability of businesses (both small and large) to achieve their core functions. The authors also examine the effect on consumers, and the broader economic consequences.

One group that has been and will continue to be affected by climate policy is business. Martin Livermore (Chapter 8) explores how European climate change policy has affected businesses and their decisions. Large businesses are able to influence the political process through lobbying, so that regulations may even give them a competitive edge. However, it is small and medium-sized enterprises which generally suffer the negative impacts of climate policy.

Businesses have responded to climate policy in various ways – by advocating the Kyoto Protocol, by investing in 'renewable' energies and by increasing their own efficiency. Some environmental campaigners have accused business of 'greenwashing' and are advocating new global rules on corporate accountability to force business to comply with their demands.

But Livermore says that business is 'an agent of change' and believes that in the next 50 years, business will turn scientific developments in the energy and transport sectors into 'huge benefits [for] society'. To do that though, requires

> [Enabling] business ... to do what it does best: innovate and produce goods and services in a manner which ensures that humanity's footprint on the earth is ever-lighter. And companies should realise that consumers are also citizens – they care about protecting the environment, and want to feel secure that the activities of business are not undermining that goal.

He warns that it is consumers and small business who will suffer if European businesses are constantly lambasted and over-regulated.

In international circles, some have speculated that at the behest of European businesses who are shackled with regulations that make them less efficient and less competitive, European countries might employ the Kyoto Protocol as a barrier to trade with countries who have not ratified the agreement, such as the USA and Australia. Julian Morris (Chapter 7) argues that there is an evolving conflict between multilateral environmental agreements, such as Kyoto, and the rules-based trading system of the World Trade Organization.

Environmental groups believe that the process of trade causes environmental problems such as global warming. They have pursued a strategy to limit trade on the basis of how goods are produced, and have been joined in coalition by businesses who face more and more stringent regulations, and who are less and less competitive as a result.

Morris argues that if the Kyoto Protocol is used to restrict trade, it will lead to poor decisions about the use of natural resources, higher consumer prices, lower economic growth, less expenditure on high-value environmental protection and lower welfare for Europeans and others.

Part Four: The Broader Context of Climate Change Policy

Climate policy has not been developed in a vacuum – it has been motivated by varying interest groups, and justified for a variety of reasons, including to avoid the threat of civilisation collapse.

Similar to other regulatory policies, climate policy has been motivated by vested interests who believe that they can gain a competitive advantage by lobbying for regulation. In other cases, these interest groups believe that they are promoting the public interest. The Kyoto Protocol has been no exception.

Bruce Yandle and Stuart Buck (Chapter 9) explain the political economy of climate change, using the analogy of a 'bootlegger and Baptist' coalition to explain negotiations on the Kyoto Protocol.

According to Yandle and Buck, during the negotiations of the Protocol, 'some nations and at least one community of nations dictated Kyoto's terms in strategic ways to enhance their positions relative to other nations.'

To the public's eye, the 'Baptists' (environmental groups) appear to be motivated purely by the public interest, even though they simultaneously promote the interests of 'bootleggers', who are 'special interest groups who are positioned to gain from regulatory enforcement and stringency or who must fend off losses that spring from proposed rules'. The authors explain:

> Day after day, newspapers and television continue to report the
> alarmist pleas of the 'Baptists' urging world leaders to 'do
> something' about global warming, but the machinations of the
> 'bootleggers' largely go unnoticed.

In Europe, climate control through the Kyoto Protocol has been accepted at face value as the appropriate solution to global warming. According to Carlo Stagnaro (Chapter 11), vested interests in Europe have an incentive to 'greatly exaggerate the risks deriving from climate change and the policies needed to address it'. Stagnaro suggests that as a result of these vested interests, Europe has a poorly formulated climate policy which will drastically reduce the GDP of European countries. Today European countries face high unemployment rates, increasing public debt and stagnant economic growth rates. Meanwhile, environmental groups suggest that we must sacrifice economic growth to fend off climate change.

Stagnaro predicts that the Kyoto Protocol might have several negative consequences in 2008–12 (the first commitment period) for average European citizens:

> Consumers would see rapid increases in living costs – food, durable goods, heating and cooling, transportation – because all energy, not just oil and gas, would be more expensive. If emissions limits were established, the cost would be passed on by businesses to consumers. Combined with the increased cost of energy, consumers would see the buying power of their salaries greatly weakened. Because economic production would greatly slow, many people might lose their jobs.

He suggests that European countries and the European Union ought to pursue policies which encourage flexibility and innovation, so that Europeans can adapt to change – whatever those changes may be. 'By rationally facing potential problems, we will avoid wasting and diverting resources from more urgent and substantial needs,' he concludes.

According to Benny Peiser (Chapter 10), the spectre of ecological apocalypse is one of the most powerful drivers of environmental gloominess and cultural pessimism. Ecological collapse caused by global warming is increasingly used to explain the collapse of past civilisations, even though there is scant evidence that that is the case.

Peiser shows that:

> Warmer periods have had a considerably benign role in social,

economic and technological progress, but global cooling and cold spells have been largely detrimental to societies.

Though people may be tempted to worry about societal collapse because of global warming, Peiser suggests that such worries

> ... are based on misleading analogies with agricultural societies that were especially vulnerable to environmental stress and lacked the benefits of modern technologies to cope with changes.

The Epilogue explains some of the differences between 'climate control' and adaptation. It discusses, briefly, how global warming and 'climate control' became conventional wisdom amongst the public. However, there is a strong need to examine the trade-offs of different courses of action, and to allow our priorities to govern our actions and responses. Finally, the Epilogue discusses how adaptation is fostered best in an institutional framework which decentralises responsibility, provides for certainty and flexibility, focuses on processes rather than outcomes, and encourages wide sharing of benefits amongst people.

Part One

Possible Impacts of Climate Change

1 Could global warming bring mosquito-borne disease to Europe?

Dr Paul Reiter

Introduction

In the 1980s, public attention on environmental issues shifted from issues of acid rain, asbestos and the ozone hole to a new concern: global warming. In the years following the negotiation of the Kyoto Protocol, numerous articles appeared in the scientific and popular press which stated that mosquito-borne diseases such as malaria, yellow fever and dengue would threaten Europe and North America if the climate continued to warm.

By the mid-1990s, the menace of these 'tropical' diseases was top of the list of nearly every account of the dangers of global warming. Interest groups used such claims to bolster their calls for urgent political actions to stop climate change:

> Global warming and the expected climate instability that
> accompanies it can have grave consequences for our health and
> well-being ... Climate restricts the range of vector-borne diseases
> (those with animal carriers), while weather affects the timing and
> intensity of outbreaks. There are strong indications that a
> disturbing change in disease patterns has begun, and that the
> global warming trend identified by the more than 2,500 scientists
> of the Intergovernmental Panel on Climate Change (IPCC) is
> contributing to these changes.[1]

This chapter explains why such pronouncements are ill-informed and misleading. The discussion is limited to malaria, but similar concepts apply to other mosquito-borne diseases.

Climate and health

The belief that hot climates are harmful to health is very old. Hippocrates described climate, seasonality and meteorological events as determinants of human illness. Deviations from the ethnic norm, as defined by the ideal of the Greek mind and body, were also attributed to excessive heat and humidity. In essence, his treatise *On Airs, Waters and Places* was an environmentalist attempt to interpret disease and race in terms of climate.

From the seventeenth century onwards, Europeans revived this notion with graphic descriptions of the 'fevers' they encountered in the tropics, often reasoning that the febrile symptoms were directly attributable to the hot climate.

In the same period, public thirst for knowledge of the newly 'discovered' lands created a major demand for travel books. Many of these were written by explorer-naturalists to fund their expeditions. To achieve sales, they fuelled the imagination of their readers with sensationalist accounts of their travels.

As a result, popular representations of the tropics became replete with accounts of the dark, the mystical, the primitive and the shocking. Illustrations of 'tropical' diseases in medical texts, often with negative connotations towards the darker races, became a significant component of this imagery. At the end of the nineteenth century, the discovery that annoying insects – mosquitoes – transmit malaria fitted well into this fear-inspiring picture.

I believe that such imagery survives today – it is reflected in the news media's coverage of disease issues, and illustrates that in our modern psyche, concepts of the *Tristes Tropiques*[2] underlie popular attitudes to 'things tropical'.

Change is a fundamental feature of climate

Climate is commonly understood to mean the 'average weather' in a given region or zone. This definition is unsatisfactory, as it implies that unlike the obvious year-to-year variations of daily weather, long-term climate is a constant. Modern climatology recognises that change is an inherent and fundamental feature of climate.[3] Just as the yearly averages of climatic elements – e.g. temperature, humidity, rainfall, wind and airborne particles – differ from one another, so too do the averages for decades, centuries, millennia and millions of years. Therefore, climatic values cannot be referred to without specifying the time span to which they refer.

For nearly three centuries the earth's climate has been in warming phase, punctuated by several periods of cooling. This was preceded by a particularly cold period, the Little Ice Age, which was itself preceded by several centuries known as the Medieval Warm Period or Little Climatic Optimum. Such changes are entirely natural, but it is widely held that, in recent years, a portion of the current warming may be attributable to human activities, particularly the burning of fossil fuels.[4]

Climate is a major parameter in all ecosystems, and has always been a fundamental factor in human settlement, economy and culture. Episodes of climate change – such as the end of the Ice Age, the drying of the Sahara, the waning of the Medieval Warm Period and the onset of the Little Ice Age – have had an important impact on human history.[5]

However, awareness of such change has remained shadowy at best, probably because the inherent time scales are beyond the span of individual human experience. By contrast, weather – the short-term condition of climate – has a much more direct and tangible impact on daily life.

Since earliest times, weather was fundamental to the success of human activities, from agriculture to seafaring, from warfare to leisure. Feast and famine, drought and flood, health and disease – all were attributable to weather events. The universal belief in weather deities, the prominence of weather events in folklore, and the ubiquitous preoccupation with weather signs and portents are evidence that an awareness of weather – particularly a fear of inclement events – has been a major feature of the human psyche throughout history.

The significance of weather has not diminished in modern society. Indeed, in the past few decades, weather awareness, particularly in the global context, has reached unprecedented levels. Weather forecasting has become an important science, fundamental to the success of agriculture, transportation, trade, tourism and virtually every other aspect of human enterprise. Weather data are collected from every corner of the globe and disseminated in digested form by government and private agencies as an aid to decision making in all walks of life. Continually updated forecasts and other information are available to the public via the popular media. Disastrous weather events from around the world are a major news feature, with detailed descriptions and graphic illustration. With this weather awareness, a new realisation of the *changeability* of climate has emerged.

The concept of 'global warming'

Current temperatures, at least in the northern hemisphere, are broadly similar to those of the Middle Ages, in the centuries before the Little Ice Age.[6] However, from the 1940s to the late 1970s, global temperatures were in decline. This gave rise to concern that particulate industrial pollutants might be exerting a global cooling effect.[7]

Since then, as climates returned to a warming mode, interest has switched to consideration of the 'greenhouse effect', a natural phenomenon by which a range of atmospheric gases trap solar radiation in the form of heat. The principal greenhouse gas is water vapour – about 2% by volume – but public attention is mainly focused on carbon dioxide (CO_2), a gas that is essential as the ultimate source of carbon for nearly all life on the planet. From the mid-nineteenth century onwards, massive clearance of forests for agriculture, followed by an exponential rise in the combustion of fossil fuels (coal, oil and gas), has resulted in a measurable increase in atmospheric CO_2, from around 0.029% in 1890 to 0.037% today.

Many climatologists agree that this 28% increase in atmospheric CO_2, together with an increase in other 'anthropogenic greenhouse gases', may be contributing to the warming trend of recent decades.[8] The extent of this contribution remains far from clear, but the mere possibility, which implies that the trend could be reversible, has given rise to spirited scientific[9] and public[10] discussion.

Human health – and mosquito-borne disease in particular – is a prominent topic in this debate.[11]

Climate and mosquito-borne disease

In nearly all mosquito species, the female obtains the protein she needs for the development of her eggs by feeding on vertebrate blood. A complex salivary secretion facilitates feeding. It is the direct injection of this fluid into the capillaries that enables several life forms – viruses, protozoa and nematode worms – to exploit mosquitoes as a means of transfer between vertebrate hosts. They do this by infecting the mosquito after they have been ingested in a blood meal, and eventually multiplying in the salivary glands, from which they can be inoculated into a new host during a later blood meal. The majority of such organisms do not appear to affect either the mosquitoes or their vertebrate hosts, but a small number are pathogens of important human and animal diseases.

The ecology, development, behaviour and survival of mosquitoes, and the transmission dynamics of the diseases they transmit, are strongly influenced by climatic factors. Temperature, rainfall and humidity are especially important, but others, such as wind and the duration of daylight, can also be significant. The same factors also play a crucial role in the survival and transmission rate of mosquito-borne pathogens.

In particular, temperature affects the pathogens' rate of multiplication in the insect. In turn, this affects the rate at which the salivary secretions become infected, and thus the likelihood of successful transmission to another host. Of course, if the development time of the pathogen exceeds the life span of the insect, transmission cannot occur. The complex interplay of all these factors determines the overall effect of climate on the local prevalence of mosquito-borne diseases.[12]

Seasonality is a key component of climate. Summer temperatures in many temperate regions are at least as high as in the warmest seasons of much of the tropics. The crucial difference is that the tropics do not have cold winters. If tropical mosquito-borne pathogens are introduced to temperate regions in the right season they can be transmitted if suitable vectors are present; in most cases they are eliminated when winter sets in.

Mosquitoes native to temperate regions have evolved strategies to survive the winter, as have the pathogens that they transmit. In the tropics, comparable adaptations are necessary for surviving in unfavourable dry periods, which can last for several years. In both cases, such adaptations impose a seasonality on transmission. For example, before eradication, the transmission season for *Plasmodium falciparum* in Italy was July to September.[13] The same three months constitute the malaria season in Mali, where the disease is still endemic.[14]

Disease models

Much of the recent speculation on the possible impacts of climate change on mosquito-borne disease has focused on rudimentary mathematical models of transmission dynamics.[15]

However, these models have a limited value for assessing the impact of long-term climate change on disease transmission.[16] They cannot predict the presence or absence of the disease, nor its prevalence in any situation, because they do not account for the parasite-rate in humans or mosquitoes, nor any of the many ecological and behavioural factors that affect the interaction of mosquitoes and humans. For example, human behaviour and cultural traits can be crucial to the transmission of the parasite. Daily activity patterns – work, rest and recreation – the location of homes in relation to mosquito breeding sites, the design of buildings, the materials used to build them, the use of screens and bed-nets, the presence of cattle as alternate hosts for the mosquitoes, and many other factors are all highly significant.

An alternative approach is to look at the past: the history of mosquito-borne diseases at different latitudes and in different climatic eras can help us to assess how climate variables relate to many other factors that affect transmission.

History of malaria in Europe

Ancient Greece and Rome

The introduction of agriculture in Europe, around 7000 BC, led to increased populations of relatively settled people, and increasingly favourable conditions for malaria transmission.[17] The extensive deforestation that began at this time may also have contributed to its prevalence, by creating additional habitat for malaria-carrying mosquitoes. Similar ecological changes in modern times have caused major increases in the prevalence of the disease.

Contemporary accounts, together with fossil and other evidence, suggest that a gradual warming and drying occurred in the Mediterranean region throughout classical times, until about AD 400.[18] Landscape studies suggest a gradual rise in sea level over this period. Around 300 BC, beech trees (genus *Fagus*) grew in Rome, the Tiber froze in winter, and snow lay for many days. However, by the first century AD the Romans considered the beech a mountain tree, and winters were definitely less severe. Over these centuries, the cultivation of the vine and olive moved gradually northwards along the Italian peninsula. The Romans were even able to introduce wine growing to Germany and Britain, and data from imports and exports suggest that Britain became self-sufficient in wine production by around AD 300. The warming trend is clearly indicated by tree-ring studies in California, so it may have been a worldwide or at least a hemispheric phenomenon.

Literary texts from that era, such as Homer's *Iliad*, include references to killing fevers at harvest time.[19] We cannot be certain that this was malaria, but later texts confirm that the disease was a significant feature in Greek life. Indeed, there is evidence that a major wave of malaria began with the flowering of Greek civilisation and transmission rates continued its increase throughout the period of the Roman Empire.[20]

Hippocrates (460–377 BC) gave exquisitely detailed descriptions of the course and relative severity of tertian vs quartan infections.[21] He also noted their association with wetlands, and even observed that enlarged spleen (often a symptom of chronic malaria infection) was particularly prevalent in people living in marshy areas. There is a wealth of evidence that malaria was common in imperial Rome.[22] Horace, Lucretius, Martial and Tacitus were among many Latin authors who mentioned the disease. The Pontine Marshes, close to the city, were notorious as a source of infection. In the second century AD, the detailed writings of Galen and Celsus on the symptoms and treatment of 'intermittent fevers' give clear evidence that three species of parasite – *P. falciparum, P. ovale* and *P. vivax* – were commonly present.[23]

The Dark Ages

Relatively little is known about climate after the Roman era during the 'Dark Ages', but there seems to have been a cooling trend from the fifth century onwards, with some severely cold winters. In AD 763–64 there was ice on the sea in the Dardanelles, and in 859–60 the sea ice on the Adriatic was strong enough to support heavy wagons. In 1010–11 it was even cold enough for ice to form on the Nile. Again, tree-ring data from California indicate that this cooling was not restricted to Europe.

Nevertheless, the armies of Visigoths, Vandals, Ostrogoths and other 'barbarians' that swept the continent had to contend with malaria, often as a major setback to their campaigns. Several popes and churchmen, including St Augustine, the first Archbishop of Canterbury, died of malaria during their journeys to Rome. Around the turn of the millennium, the armies of Otto the Great, Otto II and Henry II suffered severely from the 'Roman Fever' during their sieges of the Holy City.

The Middle Ages

The Medieval Warm Period, which reached its peak around the year 1200, coincided with major advances in technology and agriculture, and a significant increase in population throughout most of Europe. The Vikings established self-sufficient colonies, cultivating oats and barley, in northern Scandinavia, Iceland and Greenland. In the British Isles, tillage was extended to much higher altitudes than is possible today, so high that there were complaints from sheep farmers that there was too little land left for grazing. English vintners were able to maintain a flourishing production of high-quality wine, despite efforts by Bordeaux traders to restrict English exports by treaty.

The explosion of economies and culture that occurred during this warming period has been attributed, at least in part, to the beneficial impact of the warming climate. From caliphate Spain to Christian Russia, numerous medieval writers, including Dante and Chaucer, mentioned 'agues', 'intermittent fevers', 'tertians', 'quartans' and the like.[24]

Favourable temperatures and rainfall may have enhanced malaria transmission in earlier years, but Chaucer's lifetime coincided with a cooling trend that culminated in a series of severely cold winters in the first decades of the fifteenth century.[25] Much of the earlier agricultural expansion was reversed. There were many years of famine, and a large-scale abandonment of farms. Despite this cooling, malaria persisted, even in northern regions.[26]

The Little Ice Age

The first half of the sixteenth century was warm again. Temperatures were probably quite similar to those of the period 1900 to 1950. In the middle of the century, however, a remarkably sharp change occurred. After a decade or so of particularly warm years – warm enough for young people to bathe in the Rhine in January – the winter of 1564–65 was bitterly cold.[27] The next 150–200 years – dubbed the Little Ice Age – were probably the coldest era of any time since the end of the last major ice age, some 10,000 years ago.[28] Yet despite this spectacular cooling, malaria persisted throughout Europe.[29]

William Shakespeare (1564–1616) was born in the year of that first fierce winter, yet there are twelve mentions of ague in his writings. He also made several allusions to the association between swampy land and disease, and the name Sir Andrew Aguecheek presumably referred to the trembling cheeks of this ineffectual hero.

The years 1594–97 were so cold and wet that wheat harvests were a disaster, yet William Harvey (1578–1657), who wrote the first descriptions of the circulation of the blood, missed much of his final year at the University of Cambridge in 1597 because of malaria. In later years he made careful observations of malaria cases in London. The marshes in the Borough of Westminster, where the Houses of Parliament now stand, were notoriously malarial. In his treatise *On the Motion of the Heart and Blood in Animals* (1628), he described the clinical pathology of the febrile episodes, including the changes in the consistency of blood that occur in serious cases:[30]

> In tertian fever ... in the first instance ... the patient [is] short-winded, disposed to sighing, and indisposed to exertion, ... the blood [is] forced into the lungs and rendered thick. It does not pass through them (as I have myself seen in opening the bodies of those who had died in the beginning of the attack), when the pulse is always frequent, small, and occasionally irregular; but the heat increasing ... and the transit made, the whole body begins to rise in temperature, and the pulse becomes fuller and stronger. The febrile paroxysm is fully formed ...

Thomas Sydenham (1624–89), a notable physician, also lived through some of the coldest years of the era, yet made frequent reference to tertians and quartans.[31] He even remarked, 'When insects do swarm extraordinarily and when ... agues (especially quartans) appear early as about midsummer, then autumn proves very sickly.'[32]

Not all the summers of the Little Ice Age were cool. The overall mean temperature was probably at least 1°C cooler than in the twentieth century, but there also seems to have been an enhanced variability of the climate, with wide differences between clusters of up to six to eight years.

Warm summers may have contributed to this and other outbreaks, but transmission was not restricted to such years. In 1657–58, snow lay on the ground for 102 days – exceptionally cold, even with respect to the climate of the times – yet Oliver Cromwell (1599–1658) died of malaria in September 1658, just as another severe winter was setting in.

Temperatures were probably at their lowest in the period 1670 to 1700, yet it was during this very period that Robert Talbor (c.1642–81) persuaded the aristocracy of England and Europe to buy prescriptions for curing malaria that he had developed in the marshlands of Essex.[33] These were based on cinchona bark, the source of natural quinine, and earned him wealth and fame throughout Europe. In the same period, Daniel Defoe (1660–1731) described life in the Dengie marshes of Essex:[34]

> a strange decay of the [female] sex here ... it was very frequent to meet with men that had had from five to six, to fourteen or fifteen wives ... the reason ... was this: that they (the men) being bred in the marshes themselves, and seasoned to the place, did pretty well with it; but that they always went into the hilly country ... for a wife: that when they took the young lasses out of the wholesome and fresh air, they were healthy, fresh and clear, and well; but when they came out of their native aire into the marshes ... they presently changed their complexion, got an ague or two, and seldom held it above half a year, or a year at most; and then ... [the men] would go to the uplands again, and fetch another; so that marrying of wives was reckoned a kind of good farm to them.[35]

Dr Mary Dobson masterfully researched the demography, epidemiology and social impact of malaria in England in this period.[36] She found that the disease was especially prevalent in areas of brackish marshland, the preferred habitat of an effective vector, *An. atroparvus*. Data from burial records show mortality rates in 'marsh parishes' were much higher than those in upland areas, and were comparable to those in areas of stable malaria transmission in sub-Saharan Africa today.[37]

After the Little Ice Age
From the early eighteenth century until the present, temperatures have gradually returned to levels that prevailed before the mid-sixteenth century. However, the marked variability of the Little Ice Age persisted

for at least 150 years. Indeed, in the 1770s, much as is happening today, there was alarm that the climate was becoming increasingly erratic, and this prompted a new emphasis on the recording of weather variables. Some of the cold periods, particularly those between 1752 and the 1840s, were probably due to major volcanic eruptions. Whatever their causation, such episodes – accompanied by major advances of the Alpine glaciers from 1820 to 1850 – persisted until a more lasting warmth was established in the late nineteenth century.[38]

A wealth of records in the eighteenth and nineteenth centuries reveals the northern limits of malaria transmission. In the British Isles, the disease was common in most of England and in many parts of Scotland, with occasional transmission as far north as Inverness. It was endemic throughout Denmark, coastal areas of southern Norway, and much of southern Sweden[39] and Finland.[40] In Russia it was common in the Baltic provinces and eastward at similar latitudes throughout Siberia. The current average January temperature in some of these regions is less than -20°C (-4°F). Clearly, the distribution of the disease was determined by the warmth of the summers, not the coldness of the winters. The northern limit of transmission[41] was roughly defined by the present 15°C July isotherm – not the 15°C *winter* isotherm, as stated in several widely cited articles on climate change.[42]

The decline of malaria in Europe
In the second half of the nineteenth century malaria began to decline in much of northern Europe. Denmark suffered devastating epidemics until the 1860s, particularly in the countryside around Copenhagen, but thereafter transmission diminished and had essentially disappeared around the turn of the century.[43] The picture was similar in Sweden, although isolated cases were still being reported until 1939.[44]

In England, there was a gradual decrease in transmission until the 1880s, after which it dropped precipitously and became relatively rare except in a short period following World War I.[45] In Germany, transmission also diminished rapidly after the 1880s; after World War I it was mainly confined to a few marshy localities.[46] The last outbreak of locally transmitted malaria in Paris was in 1865, during the construction of the *grands boulevards*, and the disease had largely disappeared from the rest of France by the turn of the century.[47] In Switzerland, the majority of foci had disappeared by the 1890s.[48]

The decline of malaria in all these countries cannot be attributed to climate change, for it occurred during a warming phase, when temperatures were already much higher than in the Little Ice Age. However, a host of other factors can be identified:

Ecology of the landscape
'Malaria flees before the plough' is an old Italian saying, and indeed in much of Europe, improved drainage, reclamation of swampy land for cultivation and the adoption of new farming methods served to eliminate mosquito habitat.

New farm crops
New root crops, such as turnips and mangel-wurzels, were adopted as winter fodder. These enabled farmers to maintain larger numbers of animals throughout the year, thus diverting mosquitoes from feeding on humans.

New rearing practices
Selective breeding of cattle, and new introductions (e.g. the Chinese domestic pig), in combination with the new fodder crops, enabled farmers to keep large populations of stock in farm buildings rather than in open fields and woodland. These buildings provided attractive sites for adult mosquitoes to rest and feed, diverting them from human habitation.

Mechanisation
Rural populations declined as manual labour was replaced by machinery. This further reduced the availability of humans versus animals as hosts for the mosquitoes, and of humans as hosts for the parasite.

Human living conditions
New building materials and improvements in construction methods made houses more mosquito proof, especially in winter – another factor that reduced contact with the vector.

Medical care
Greater access to medical care and wider use of quinine reduced the survival rate of the malaria parasite in its human host.

Much of the decline in malaria came before recognition of the role of mosquitoes in its transmission. Thus, for most of the region, deliberate mosquito control played little or no role in its eventual elimination.

The persistence of malaria in the USSR
In countries where profound changes in crop production and stock rearing were absent, malaria did not decline.

In Russia, from the Black Sea to Siberia, major epidemics occurred throughout the nineteenth century, and the disease remained one of the principal public health problems for the entire first half of the twentieth century.[49] In 1900, annual incidence in military garrisons was 6.6 per 1,000 in St Petersburg, 31.0 per 1,000 in Moscow, and several hundred per 1,000 in the more southerly provinces. Mean annual incidence from 1900 to 1904 was 3,285,820, but by the period 1933 to 1937, it had risen to 7,567,348.

Some of this increase can be attributed to more effective reporting, but there is no doubt that the disease became much more prevalent after World War I and the Revolution. In the 1920s – in the wake of massive social and economic disruption, two years of severe drought, and a year of widespread flooding – a pandemic swept through the entire Soviet Union.[50] Official figures for 1923–25 listed 16.5 million cases, of which not less than 600,000 were fatal. Tens of thousands of infections, many caused by *P. falciparum*, occurred as far north as the Arctic seaport of Archangel (61° 30'N).

The Soviet government appeared to make some headway against the disease in the 1930s, mainly by drainage schemes, afforestation, and naturalistic methods such as the use of mosquito-eating fish. World War II interrupted these efforts, and transmission soared, particularly in the Ukraine, Byelorussia and other occupied areas. Finally, in 1951, a huge multi-faceted anti-malaria campaign was initiated. It involved widespread use of DDT and other residual insecticides, antimalarial therapy, land reclamation, water management, public health education and many other approaches. This mammoth effort finally brought about a dramatic reduction of transmission, so that by the mid-1950s the national annual incidence was below 1 per 10,000.[51]

Until the collectivisation of farmland that began in the winter of 1929–30, the Soviet Union was largely unaffected by the agricultural revolution. By 1936, all farming was essentially in government hands, but in protest, many peasants had slaughtered their horses and live-stock, and destroyed their equipment. These events ran counter to many of the changes that had reduced transmission in much of Europe. In neighbouring Poland and Finland, farming was also less advanced than in much of the rest of northern Europe, and malaria continued to be a problem, but slow modernisation probably contributed to the steady downward trend in cases.

The contrast between the devastation caused by malaria in the Soviet Union until the 1950s and its quiet withdrawal from other countries at similar latitudes in the previous century is a vivid illustration of the importance of non-climatic factors in transmission.

The persistence of malaria in southern Europe
Malaria remained highly prevalent in much of Mediterranean Europe, the Balkans and the countries bordering the Black Sea until after World War II.[52] A number of effective vector species, an abundance of prolific mosquito breeding sites, the warm climate and the long summer season were all highly conducive to transmission.

In addition, much of the region was relatively unaffected by the environmental changes associated with modern agriculture. Part of this lack of change can be attributed to the disease itself, for poverty and lack of progress characterised many of the highly malarious regions. In northern Italy, for example, much of Piedmont and Lombardy was free of transmission. By contrast, large portions of the rest of the country, particularly in Sardinia, Calabria and Sicily, remained virtually uncultivated until the 1950s, at least in part because of the ravages of the disease. The same was true for major regions in Spain, Greece, Romania and Bulgaria.[53]

The final elimination of malaria from Europe
Until the end of World War II, the only effective approach to mosquito control was to target the breeding sites – by environmental modifications such as drainage and landfill, and by the application of insecticidal oils or chemicals. These methods were costly, so they were mainly applied to urban centres and other areas of high economic importance. The advent of DDT revolutionised malaria control.[54] It enabled cheap, safe, effective treatments to be targeted at the site where most infections occur – in the home.

The principal treatment method was to apply 2 gm/m^2 to indoor surfaces once every six months. Mosquitoes were killed by contact when they alighted on the treated surfaces. Initial efforts in Italy, Cyprus and Greece were so successful that a decision was made to eradicate the disease from all of Europe.[55]

The campaign was based on a careful application of scientific principles, meticulous planning, efficient administration, generous financing and continuous emphasis on evaluation. It was orchestrated by several international agencies, particularly the World Health Organization (WHO) and the United Nations Children's Fund (UNICEF), as well as numerous national bodies, including the US Public Health Service. The International Health Division of the Rockefeller Foundation also provided generous financial and technical support. By 1961, eradication had already been achieved in many countries. The entire continent was finally declared free of endemic malaria in 1975.[56] One of the last countries to be declared free of malaria was Holland.

Holland: an illustration of the complexity of malaria transmission

The persistence of malaria in Holland, a country that has held a central position in the economic life of Western Europe since the Middle Ages, is a good illustration of how local conditions make malaria transmission extremely complex.

In the nineteenth century, despite great progress in drainage and cultivation, the heavily populated 'Low Countries' were the most malarious region of northern Europe. During the Napoleonic Wars, the failure of the British Walcheren expedition (1809) was blamed on malaria, after 4,000 troops died of fever. Severe outbreaks, also attributed to the disease, occurred in 1826, 1834 and 1846, with many thousands of deaths.[57]

The dominant vector, *Anopheles maculipennis*, was present throughout the country. However, the disease was particularly prevalent in areas that had been reclaimed from the sea, especially in Noord Holland, a province that includes the cities of Haarlem and Amsterdam. Moreover, a major peak of incidence occurred in the springtime, rather than summer or autumn.

The unravelling of this puzzle was one of the classic triumphs of medical entomology.[58] Investigation revealed that *An. maculipennis* was not one, but several 'sibling' species:[59] *An. atroparvus*, which lays its eggs in brackish water; *An. messeae*, which prefers fresh water; and

An. maculipennis sensu strictu, which occurred inland, in non-malarious areas.

In the laboratory, *An. messeae* and *An. atroparvus* were excellent malaria vectors. In the field, the picture was very different. Both species preferred to feed on domestic animals rather than humans, but whereas *An. atroparvus* rested in stables, *An. messeae* preferred uninhabited sheds and other unheated outhouses. In the autumn, *An. messeae* built up a fat body that allowed it to hibernate. By contrast, *An. atroparvus* remained semi-active, feeding at regular intervals throughout the winter. Although most of its meals were taken on farm animals, it occasionally wandered into adjacent human dwellings. If persons infected with the malaria pathogen were present, the mosquito acquired the infection.

Thus, as in many tropical countries, transmission in Holland occurred at all times of the year, despite winter temperatures that could dip below -20°C. However, the mosquito's ovaries did not develop eggs until the advent of spring, a condition known as *gonotrophic dissociation*. An additional twist to the story was that the local strain of *P. vivax* had a particularly long incubation period, so persons infected in the autumn and winter only showed symptoms of illness in the spring, after the first new brood of mosquitoes had emerged.[60]

In 1932 a dyke was built to enclose the Zuider Zee, a vast area of brackish water to the east of Noord Holland. The accumulation of river water behind this dyke caused a gradual decrease in the salinity of the surrounding land, greatly reducing the larval habitat of *An. atroparvus*.[61] In addition, there were major changes in the living conditions of humans. New farmhouses were less intimately associated with cattle sheds and stables, and their structure and heating technology made them much less hospitable to overwintering mosquitoes.[62]

Thus, although the disease was finally eradicated by routine DDT applications and the administration of anti-malarial drugs, the ecology, physiology, behaviour and survival of the vector, plus the interaction of the pathogen with both vector and host, all contributed to its demise.

Could malaria return to Europe?

In much of Europe, changes in lifestyles and living conditions were the most important factor in the elimination of malaria. Even in countries where these factors were less dominant, eradication of the disease did not require total elimination of the vector. Residual treatments were effective because they reduced the life span of the adult insect, reducing the probability of transmission and eventually leading to the elimination of the parasite. Thus, the malaria-carrying mosquitoes are still present in the brackish waters of England, the rice fields of Italy, the ponds of Poland, the forest pools of Finland and the riverine swamps of Russia.

Advances in agriculture and improvements in living standards have limited the mosquito populations and reduced their contact with humans in many regions, but this is not always the case. For example, recent surveys show that in large areas of Italy, *Anopheles* mosquito populations have returned to levels not seen since before the DDT era. In entomological terms, the infestations are comparable to those in areas of Africa that have extremely high transmission rates.

Nevertheless, the malariogenic potential of Italy is considered to be very low, and re-establishment of malaria is judged unlikely unless living standards deteriorate drastically.[63] Moreover, if the present warming trend continues, human strategies to avoid warmer temperatures – particularly indoor living and air conditioning – are likely to become more widespread.

Of course, this does not mean the disease will be entirely absent. International travel and population movement will facilitate introductions from other parts of the world. For example, in 1997 the WHO recorded 12,328 cases of imported malaria in the European region. Such cases occasionally lead to summertime transmission,[64] recently reported as far north as Toronto and Berlin. However, in all the wealthier countries, such outbreaks are likely to be small, easily contained and confined to a limited geographic area.

The same may not be true of less affluent regions. Rapid economic decline, combined with political instability, has already brought back epidemic typhus, diphtheria and other infectious diseases to several countries of the former Soviet Union.

In the 1990s, epidemic malaria has made a dramatic reappearance in Armenia, Azerbaijan, Tajikistan, Turkey and Turkmenistan. Cases have also been reported from Dagestan (Russian Federation), Georgia, Kazakhstan, Kyrgizstan and Uzbekistan.[65] It is quite possible that the disease could spread northward into Russia and westward around the Black Sea. The 1999 conflict in the Balkan states was in the same region where hundreds of thousands were infected with malaria during World War I. Endemic transmission in such areas could be significant if the parasite were to be reintroduced. Climate change might augment this possibility, particularly at high latitudes (e.g. in Siberia) although low stability should facilitate control.

Perspective on the global debate

The history of malaria in Europe, especially during periods when the climate was much colder than it is today, contradicts the popular notion that the disease is restricted to the tropics, yet the authors of many influential publications on the impacts of climate change evidently share this misconception.

For example, in 1995, the International Panel on Climate Change (IPCC) confidently forecasted that malaria and other mosquito-borne diseases would move from the tropics into temperate regions.[66] Similarly, in 1997, the Environmental Protection Agency (EPA) of the US government stated that in the 21st century there would be: 'an increase from approximately 45 percent to 60 percent in the proportion of the world's population living within the potential zone for malaria transmission.' In their estimate, this could result in 50–80 million additional cases annually.[67] Several other publications included maps that showed the future range of the disease extending into southern Europe.

In the past two years the subject has been treated more cautiously, but the intuitive assumptions and predictions of the 'spread' of malaria from the tropics into temperate regions still persist. Thus, in its *Third Assessment Report*, the IPCC repeats the assertion that, 'the geographic range of malaria is limited to the tropics and subtropics',[68] and the EPA continues to state that, 'global warming may also increase the risk of ... infectious diseases ... that only appear in warm areas. Diseases such as malaria could become more prevalent if warmer temperatures enabled [mosquitoes] to become established farther north.'[69]

Environmental activists quote these official statements and add warnings that are even more graphic. For example, the World Wildlife Fund quotes the IPCC, followed by a statement that 'malaria generally extends only to places where the minimum winter temperatures reach no lower than 16°C.' It even asserts that 'small outbreaks now occurring north and south of tropical regions are consistent with model projections' and supports this with claims of local transmission of malaria in the US and Canada 'during particularly hot, humid periods'.[70] The outbreaks it refers to are all associated with imported cases and occurred in regions where malaria was once common.

I believe that such misinformation should be treated seriously. There is much talk of efforts to improve the health of poorer nations. At the same time, erroneous concepts of mosquito-borne disease are used as an argument to spend colossal amounts of scarce resources to 'halt' global warming, even as climate experts confess that the true contribution of human activities to the present warming trend is uncertain.

History is replete with bizarre decisions based on superstition and misconception. There is a tendency to assume that modern science is proof against such error. While it is true that public policy is increasingly driven by science, it is also true that much of science is nurtured by public policy.

The story of malaria in Europe is widely known and readily accessible in any good library. Nevertheless, uninformed predictions on the spread of this and other vector-borne diseases to temperate areas are commonplace – even in the scientific literature – and are widely quoted in public discussion of national and international policy on global warming.

In my opinion, these predictions are sustained by (1) Hippocratic concepts of the association between climate and disease; (2) a fear of things tropical; (3) a disregard for the past; and (4) a necessity to simplify concepts for public consumption. The truth, as we have seen, is that the natural history of malaria is complex, and the interplay of climate, ecology, vector biology and many other factors defies simplistic analysis.

The sad fact is that there is little we scientists can do to challenge the campaign of misinformation. None of us denies that temperature *is* a factor in the transmission of mosquito-borne diseases, and that transmission *may* be affected if the world's climate continues to warm. But it is immoral for political activists to mislead the public by attributing the recent resurgence of these diseases to climate change, particularly in Africa. The true reasons are far more complex, and the principal determinants are politics, economics, and human activities. A creative and organised application of resources to correct the situation is urgently needed, regardless of future climate.

This article is a modified version of a French article originally published in 'Annales de l'Institut Pasteur/actualités: Changements climatiques, maladies infectieuses et allergiques', pp. 63–89. (Elsevier SAS.)

2 Barriers to barriers: why environmental precaution has delayed mobile floodgates to protect Venice

Dominic Standish

Venice and the climate change debate

Venice has been implicated as one of the victims of global warming. The news media have sensationalised Venice 'sinking' as a result of rising sea levels. 'A global threat laps at the gates of Venice,' cried the headline to a climate change feature on the front page of the *International Herald Tribune*.[1]

The Italian World Wildlife Fund (WWF) and conservationists Italia Nostra (Our Italy) promote the idea that Venice may soon be inundated because it will be subjected to rising sea levels from climate change. However, the Italian Green Party, working with these environmental groups, has also delayed Project MOSE, a system of 79 mobile barriers that would block the high tides that cause flooding.

They object that Project MOSE will damage the ecosystem of Venice and violate nature. WWF and Italia Nostra have jointly demanded the 'renaturalisation' of the lagoon.[2] Environmentalists have caused a false dilemma for Venice, for the question is not 'to preserve or not to preserve', but how Venetians can save their historical city from sinking and from flooding. This chapter will examine a specific case of how the precautionary principle informs environmental risk decisions in the context of climate change.

Venice's unnatural history

The formation of Venice was one of the most unnatural acts imaginable, instigated because a group of people sought to use its habitat to their advantage. Why would anyone try to build a city in a lagoon? As John Julius Norwich has eloquently described, its creators were the most intelligent people from the nearby mainland, who were fleeing barbarian invaders:

> Thus it was that the wisest came to the islands of the lagoon.
> There, they believed, these savages from landlocked central
> Europe, lacking both ships and knowledge of the sea, would –
> with any luck – ignore them, turning their attention instead to the
> richer and far more tempting prizes on the mainland. They were
> right.[3]

Over time, they found that the lagoon's waters offered additional advantages, such as fish and salt. But constructing communities in a lagoon required extensive intervention in nature. Man-made land and dwellings were constructed in the lagoon's islands that offered settlements. Land needs were greater than nature provided and foundations were extended into the water.

In medieval times remarkable feats were achieved to pile the lagoon bottom with millions of alder tree supports on which a thick layer of Istrian Limestone slabs provided a flat and stable surface for building foundations. The top slabs were positioned at the water surface, thus protecting the structures from its impacts. Century on century created modern-day Venice with its 100 island units separated and connected by canals and bridges.[4]

Venetian society has constantly intervened in the lagoon for human development and to maintain the very existence of the lagoon. For instance, the Brenta and Sile Rivers that formerly flowed into the lagoon were diverted, as was the Po River to the south, a process that was completed in 1604.[5] Other major projects to maintain the lagoon have included building defensive walls against the Adriatic Sea in 1744–83. These were considerable feats given that much of the work was done by hand.

Why is Venice 'sinking'?

The term 'sinking' confuses a number of phenomena that have caused flooding in Venice. Firstly, it is important to distinguish between the general rise in relative sea level (RSL) and the exceptional high tides, even though both contribute towards flooding. The exceptional high tides are the product of many factors, including rain, winds known as the *scirocco* and *bora* and storm surges. Storm surges in the Adriatic Sea do not appear to be related to global warming,[6] even though their frequency is increasing.[7] The astronomical causes of high tides can easily be predicted[8] and storm surges can be forecast several days in advance.[9] This is vital as the narrowness of the Adriatic makes its tides higher than in other parts of the Mediterranean Sea. Flooding has also been exacerbated by the deepening of the lagoon's channels and outlets to the sea for shipping, which has increased the speed of tide currents in the lagoon.[10]

Between 1897 and 1983, the RSL in Venice rose 23 cm, according to measurements by the Italian National Research Council; 12 cm of the 23 cm RSL rise was due to subsidence and 11 cm was caused by rising sea levels.[11] Current subsidence, or lowering of the land level, is mainly due to the self-weight of the city[12] and erosion of the lagoon's bed.[13] Venice's sea levels have risen for many reasons, including the shifting of geological plates[14] and climate change.

The principal reason for subsidence in Venice during the twentieth century was the extraction of groundwater from wells for the industrial complex at Marghera in the lagoon, especially between 1930 and 1970. Ghetti has explained how Venice's natural geological rate of 0.4 mm subsidence a year accelerated to 1.8 mm a year from 1930, and by 1950 it was 8 mm a year.[15] It became worse between 1968 and 1969 when there was sinking of 17 mm at Marghera and 12 mm in Venice.[16]

Marghera's oil refinery was a serious mistake in terms of Venice's subsidence. But there was a positive response that stopped this abnormal subsidence: the Italian government ordered the closure of a large number of wells for Marghera's industrial zone and supplied it from surface watercourses after 1970. Greenpeace has run poison tours to parts of the lagoon they claim are 'toxic' and has continued to organise stunts at Marghera in its campaign against it. But industry and climate changes are not the only causes of flooding in Venice. As chronicles show, flooding has been plaguing the city for centuries[17] and the rise in RSL has recently slowed. Now it is important to focus on the current and future causes of flooding.

There is a general consensus that subsidence can be predicted fairly accurately at 4 cm over the next 100 years,[18] the planned lifespan of Project MOSE. The contemporary controversy is over the impact of climate change on the RSL for Venice.

Flooding in Venice

Venice is now flooded roughly 43 times a year, compared with seven times a year at the start of the twentieth century,[19] causing damage to buildings and monuments. In 1996 the city was flooded on 100 occasions and in 1966 there were exceptional floods of almost 2 m above the sea-level tide meter at Punta della Salute.[20]

By 1973, the Italian Parliament had passed Special Law 171 concerning Venice, and the idea of constructing mobile barriers to protect the city was legally recognised. The preliminary design of Project MOSE was completed in 1992. It was then approved by the Venice Water Authority Technical Committee (1992) and the Higher Council of Public Works (1994). However, it was not until 3 April 2003 that a system of mobile barriers was given definitive approval, with estimated completion in 2011.

The reasons for such a long delay in protecting Venice are varied and have changed over time. Some have blamed Italy's unstable political culture, the rigid bureaucracy and the system of political favours that have made continuity in public works difficult.[21] There is also the cost of the mobile barriers, currently estimated at US$3.5 billion. While political culture and costs have been influential, environmental concerns have become instrumental in delaying the mobile barriers since the mid-1990s.

Project MOSE

MOSE is an acronym for *Modulo Sperimentale Elettromeccanico* (Experimental Electromechanical Module). It is the system of 79 mobile barriers designed to protect the three entrances (at Malamocco, Lido and Chioggia) to the lagoon that surrounds Venice. The barriers would remain below the surface of the sea until high tides and flooding of the city are predicted, especially when storm surges are expected.[22] Then Project MOSE's mobile gates will rise up to block high tides 110 cm above the Punta della Salute tide meter. A forecasting system will predict when the 110-cm flood limit will be exceeded, closing the gates at 55 cm six hours before 110 cm would be reached.[23]

Project MOSE has been designed and is being constructed by Consorzio Venezia Nuova (New Venice Consortium) (CVN), a private pool of Italian building contractors established by the Italian state to oversee the design, construction and monitoring of work to safeguard the lagoon. CVN acts as a contractor for the Ministry of Infrastructure and Transport and the Venice Water Authority.

For Giancarlo Galan, President of the Veneto region, Project MOSE is 'historic'. As the construction began, Galan proclaimed 'the safeguarding of the city is finally starting to be carried out so there is hope for Venice.'[24]

Additional and alternative anti-flooding proposals

Project MOSE's system of mobile barriers was chosen after considering many other mechanisms to separate the sea from the lagoon and after over a decade of testing. But its opponents, such as WWF and Italia Nostra, are still proposing alternatives that will have a minor environmental impact. The problem is that protecting Venice from flooding requires transforming the relationship of the city to the sea with significant impact.

The most popular alternative is simply to raise low-lying areas of Venice by 120 cm, much more than is currently planned. An International College of Experts, established in 1996 by the Italian government to evaluate Project MOSE, is sceptical of this alternative:

> Such a measure would reduce the frequency of flooding under the present sea-level conditions down to about once a year. In practice, however, it would be difficult to implement, costly and time consuming. This alternative is as expensive as the mobile gates, would take 60 years or more to be completed, and would leave the city exposed for a long time to come. Since it would not protect against exceptional floods above 120 centimetres, it should be considered as only a partial solution.[25]

Many non-governmental organisations (NGOs)[26] and opponents of
MOSE have called for alternative forms of intervention that they believe are more beneficial for the lagoon's ecosystem:

> These 'alternatives' include changing the geometric configuration
> of the breakwaters, filling the so-called *'Petroli'* (oil-tankers)
> channel, or reducing the depth of the sea bed at the lagoon
> openings. Although the utility of such measures is not entirely
> clear, their costs are prohibitive, and potential negative
> consequences are disastrous (to the point that they could in some
> cases be defined as *extravagant* measures).[27]

The Report by the International College of Experts also stated that
'Such measures cannot be regarded as valid alternatives. However,
some of them could be complementary to the mobile gates.'[28]

It is often assumed that diffuse measures will benefit the lagoon's
ecosystem. But the International College of Experts ascertained that
diffuse measures would have both beneficial and negative environmental impacts.[29]

CVN and other bodies have already improved the lagoon substantially in recent years utilising diffuse measures, including improving
sea defences and water quality, and re-establishing the lagoon's morphology.[30] CVN helped to implement a programme that improved the
ecosystem of the Palude della Rosa marsh in the northern lagoon.[31] A
report on bird habitats concluded 'Various measures planned and implemented by Consorzio Venezia Nuova along the coastline could
have positive repercussions on the composition of the fauna present in
these habitats.'[32]

In addition, CVN and other organisations have begun to implement *'insulae'* (interior island infrastructure) defence measures. They
have elevated paved surfaces outside and even inside buildings, and
they have raised the height of seawalls surrounding the islands and
along canal banks and directly lifted structures. These measures will
protect the lagoon's towns, villages and Venice when tides rise up to
110 cm on the Punta della Salute tide meter. The gates from Project
MOSE will protect Venice from tides above 110 cm.

These diffuse and *insulae* operations will complement Project
MOSE, but will also have a minimal impact on reducing the flooding.
In combination with the 'extravagant' measures called for by some

NGOs, it has been estimated that they 'would generally contribute to a reduction in the high water level in Venice of less than 6 cm (and the single and/or cumulative effect would be even smaller in the case of steep tidal flows)'.[33]

The CVN company has based its work on environmentally friendly policies with 'environmental auditing-reporting based on eco-indicators' and the 'continuous re-orientation of engineering towards sustainability'.[34] For example, prefabricated systems were chosen to limit the impact on the local environment during construction of the gates, although some negative environmental consequences are inevitable during construction.[35] The operation of Venice's port has also integrated environmental precaution by involving NGOs in identifying and planning to minimise environmental risks.[36]

Green opposition to Project MOSE

Environmental opposition to Project MOSE has taken various forms as political conditions have changed. Although the Greens have enjoyed little success in general elections, it would be a mistake to ignore how they have influenced public policy on issues such as the mobile barriers.

The ability of Green Party members and their associates in environmental NGOs to challenge the mobile barriers was transformed by the inclusion of the Green Party in government for the first time in 1996.

By 1998, the Italian Environment Ministry, dominated by the Green Party and influenced by environmental NGOs, had produced a 400-page report criticising Project MOSE through the Ministry's National Committee of Environmental Evaluation. The report asserted that other anti-flooding measures should be pursued before a revised version of the mobile barriers 'could be reconsidered once basic work to re-establish the general health of the lagoon and the city has been undertaken and its effects on their vulnerability to flooding taken into account'.[37]

Based on this report, the Environment Minister and leading Green Party member, Edo Ronchi, issued a decree halting Project MOSE at the end of 1998 until the Regional Administrative Tribunal for the Veneto (TAR) ruled this decree invalid for technical reasons in 2000. Then the Green Party made abandoning Project MOSE a condition for its participation in the 2000 government led by Giuliano Amato.[38]

The Green Party was not represented in the next government elected in 2001 and headed by Silvio Berlusconi. As prime minister, Berlusconi chairs the Committee for Policy, Coordination' and Control (Comitatone) that is responsible for coordinating all measures to safeguard Venice and the lagoon. The Comitatone unanimously approved the final design stage of Project MOSE on 6 December 2001. The first financial instalment of €450 million for Project MOSE's development from 2002–4 was approved on 29 November 2002. Berlusconi led the inauguration ceremony for Project MOSE on 14 May 2003. 'This is not a good day for Venice, or Italy,' was the response of Eduardo Zanchini, who monitors the mobile barriers for Italy's leading environmental group, Legambiente.[39] The Green Party exerts considerable influence within Venice's City Council that has a long history of challenging the central government's attempts to implement MOSE. Venice's City Council contributed towards stalling definitive approval for Project MOSE at the government's Comitatone meetings on 4 and 25 February 2003. When Project MOSE was given the final go-ahead on 3 April 2003, Venice's City Council put up eleven conditions, including plans to raise pavements and protect lower-lying parts of Venice from floods, and testing some alternatives to Project MOSE. Venice's Mayor, Paolo Costa, managed to prevent outright revolt from Greens in Venice's City Council by presenting the eleven conditions for 'the equilibrium of the lagoon's morphology'.[40]

The government's Comitatone Committee was under no obligation to consider these conditions to give definitive approval to Project MOSE. Acceptance of the eleven conditions indicated that even Berlusconi's government has capitulated to environmental precaution. This government, with its exceptionally large parliamentary majority, was under no pressure to give in to Green Party members with their weak parliamentary representation. These conditions and this agreement were satisfactory 'from an environmental point of view,' said the Environment Minister, Altero Matteoli.[41]

Most NGOs are still challenging Project MOSE. Court actions are central to their opposition. In 1998, Italia Nostra attempted to stop the floodgates by taking a case to the European Commission that claimed the CVN's control of the project was an illegal monopoly. In 2002, much to the annoyance of environmentalists, the Commission closed the case, allowing CVN to continue Project MOSE providing parts of it were put out to bidding.

Earlier this year, WWF and Italia Nostra worked closely with Venice's City Council and the Veneto Regional Council to suspend work already started by CVN at Venice's Malamocco entrance to the lagoon. They successfully appealed to the TAR Tribunal in February 2003.[42]

Undoubtedly, there will be many further battles with environmentalists before Project MOSE is completed.[43] In Italy's turbulent political system, the tide could easily turn against the mobile barriers. The leader of the government opposition coalition is former Green Party member Francesco Rutelli. A change of government could mean that funds for Project MOSE dry up.

Global warming and Venice

Despite the scientific uncertainty regarding climate change, the media have contributed to the linkage of global warming and Venice 'sinking'. This perception is supported by many experts who also argue that climate change means Project MOSE is useless. 'Floodgates "won't save Venice"'[44] and 'Venice flood barriers scheme "will soon be obsolete"'[45] were the titles of articles for BBC Online and the UK's *Independent* newspaper after Paolo Pirazzoli published an article for a journal of the American Geophysical Union. Pirazzoli, a marine geophysicist at the French National Centre for Scientific Research, believes that Project MOSE does not take global warming into account: 'The weakness in the project can be explained by the fact that the system was officially put forward in 1981 and has not been subsequently adapted to the predictions of greenhouse gas buildup-related sea-level rise which have been foreseen since 1982.'[46]

To delay Project MOSE, interest groups have used studies by experts who assert that global warming will make the project deficient. For example, *Environment Magazine* has championed the work of the American archaeologist Albert Ammerman as 'the gates' most prominent critic'.[47] Ammerman and his colleague, Charles McClennen from Colgate University (New York), contend that their campaign against Project MOSE is based on their archaeological insights in an article in the journal *Antiquity*.[48]

However, it was an article by Ammerman and McClennen in the journal *Science* that provoked a reaction beyond the scientific community.[49] Ammerman and McClennen asserted that the Environmental

Impact Study's scenarios for RSL were invalid because they were based on short-term tide-gauge data.[50] But the International College of Experts analysed the EIS scenarios and endorsed them.[51]

Ammerman and McClennen estimated Venice's RSL rise:

> If we start with the average, long-term rise in RSL as a baseline (13 cm per century), add a safety margin (4 cm per century), and make a minimal allowance for global warming (13 cm over the next 100 years), a value of 30 cm is obtained for a new low projection of the rise in RSL. The 'worst-case' scenario (high estimate) would be on the order of 100 cm.[52]

But the data presented here does not justify the authors' projections. They correctly highlighted the need for measurements over a long period of time to establish reliable trends. Their own archaeological work has led them to their estimates from AD 400 through to 1900 for RSL trends. This is used for the first part (13 cm) of their forecast for RSL rise during the 21st century.

This estimate is rather low considering the RSL rise was 23 cm during the twentieth century, as the authors acknowledge.[53] The additional 13-cm estimate 'for global warming' over the next 100 years is not based on their archaeological work, nor is the 4 cm 'for safety'. So only 13 cm out of 30 cm of their low estimate for RSL rise is justified by their archaeological findings, and they did not explain how they arrived at their 'high estimate' of 100 cm.

Ammerman and McClennen based their predictions for global warming on forecasts by the IPCC. Indeed, in the *Science* article they wrote: 'If the new Intergovernmental Panel on Climate Change (IPCC) report that is forthcoming sustains its previous position on global warming, then the handwriting could be on the wall regarding the project.'[54]

In fact, the IPCC's Third Assessment Report in 2001 published after the *Science* article revised its estimated range of total sea-level rise until 2100 downwards from a range of 38 cm to 55 cm, to a range of 31 cm to 49 cm.

While opponents of Project MOSE have published papers arguing the project will soon be useless, the substantive studies on Project MOSE have been positive. The EIS gave Project MOSE overall approval. The International College of Experts, assembled for an objective assessment of the EIS and other studies, was positive about the project. Rafael Bras of the Massachusetts Institute of Technology (MIT), who disputes predictions that high sea-level rises will make the gates redundant, led this team of international experts: 'The bottom line is that the gates work ... To argue that the design of the barriers did not consider sea-level rise is just wrong ... The barriers, as designed, separate the lagoon from the sea in an effective, efficient and flexible way, considering present and foreseeable scenarios.'[55]

Debates will continue about climate change and global warming because sea-level rises are very difficult to predict in assessing how to protect Venice from flooding.

How will rising sea levels affect Venice?

The debate about Project MOSE and rising sea levels focuses on the extent of predicted gate closures. A large number of gate closures could cause potential problems for shipping and pollution in the lagoon. Projections for gate closures inevitably depend on forecasts for the RSL, because higher rises will increase the number of occasions when the gates will block the sea from the lagoon.

According to Brotto and Gentilomo, a rise of 20 cm to 30 cm in RSL would generate a negative number of gate closures. This would have an impact on ships, which would be restricted from entering the lagoon.[56] The current sea level means the gates would be closed for 80 hours a year[57] disrupting 2.3% of sea traffic.[58] A 20-cm RSL rise would mean closures for 500 hours a year[59] disrupting 13.3% of traffic.[60] Each gate closure will last approximately 4.5 hours.

Ammerman and McClennen believe that their low projection for RSL rise (30 cm over the next 100 years) could produce 94 to 150 gates closures a year.[61] In addition, the Environmental Impact Study considered scenarios for gate closures over the next 50 years. It concluded that under the worst-case scenario for sea-level rise of 20 cm by 2050, the number of gate closures would increase from twelve per year under current conditions to 70.[62]

If the RSL does rise to produce a high number of gate closures, there are a number of possible positive responses. Forecasting gate closures and better planning and communication about when to pass into the lagoon could reduce disruptions to sea traffic. The diffuse measures that are vital to complement Project MOSE could also

minimise the need for gate closures. Raising vulnerable areas of
Venice by 120 cm would keep the number of gate closures to twelve
per year with a 20-cm RSL rise by 2050, according to the International College of Experts.[63]

Opponents of Project MOSE also believe that gate closures will
cause more pollution. Ammerman and McClennen have described the
seasonal nature of Venice's high tides, and they predicted frequent closures during the winter months of October through January.[64] They
suggested: 'As such a high concentration of gate closures will limit the
circulation of water that is essential to biological life in the lagoon,
this could have negative impacts on levels of water pollution and the
ecology of the lagoon.'[65]

It is true that gate closures are more likely during the winter. But
pollution is a much greater problem in Venice during the summer than
the winter, as examined by Cecconi (as head of the CVN engineering
department).[66] So a concentration of gate closures during the winter
may actually be more beneficial than if the gate closures were spread
out over the year.

Closing Project MOSE's gates could also have a positive impact on
the lagoon, reducing pollution by flushing the whole lagoon and decreasing the presence of pollutants. Reducing the ebb and flow could
also minimise the silting of canals and the loss of sediment from the lagoon.[67] A report by experts connected with the CVN company explained how the mobile gates could be used to reduce pollution and
benefit marine life:

> Once storm surge gates have been installed, they can also be used
> to induce lagoon flushing. By preventing water leaving through
> the central tidal opening, a net circulation flux through this
> opening of 2000m(3)/s will result. This will flush the tidal flats
> south of Venice and immediate benefits can be expected in terms
> of the oxygen content during periods of fast *Ulva* biomass
> decay.[68]

The International College of Experts found that with the current sea
level the gates would need to be closed on twelve occasions a year and
that this would have 'negligible' impact on the lagoon's environment.
With a possible sea-level rise of 10 cm between 2030 and 2100, the
gates would block the sea from the lagoon 30 times a year: 'By this

time, some effect on the natural system of the lagoon may become measurable, although it is expected to be small.'[69]

The experts did foresee a 'measurable' effect on the environment if the sea level rises by 20 cm leading to 70 gate closures a year, which could happen by 2050 with a pessimistic scenario. However, their report points out that by the middle of the 21st century, we will have the benefit of increased knowledge and experience to inform decisions about how to protect Venice.[70]

For now, Project MOSE plus the diffuse/*insulae* measures appears to be the best way to protect Venice. Even if we accept the pessimistic predictions for RSL rise and gate closures, the potential problems regarding shipping and pollution seem manageable.

If climate change brings very significant sea-level rises, Project MOSE would not necessarily be redundant. Additional measures could be carried out with the mobile barriers continuing to provide some protection. For example, local defences could be raised by a further 20–30 cm beyond current allowances, shipping locks could be constructed, or the lagoon could be separated into several basins[71] or two parts (sub basins).[72]

This adaptive approach to Project MOSE and the complementary measures is the best way to protect Venice against immediate and future RSL rise. The gates are mobile, so they can flexibly respond to separate the lagoon from the sea as conditions change.

Environmental risks and the precautionary principle

Italian environmental groups have based their opposition to Project MOSE on a profound aversion to risk. It is claimed that Venice's future is being placed in jeopardy by reckless gambling. Gaetano Benedetto of WWF Italy greeted the approval of Project MOSE's final design stage by stating: 'Today the city's destiny rests on a pretentious, costly and environmentally harmful technological gamble.'[73]

In fact, standard risk assessments of Project MOSE have been conducted by engineers to test its safety. A committee of international experts from leading engineering companies approved the design of the project in 1993. But adopting the precautionary principle implies that 'technology ought to be severely restricted if not banned, unless it can be proven to be absolutely safe.'[74]

Tony Zamparutti, a member of the environmentalists' committee

'Save Venice with its Lagoon', recently wrote in *The Ecologist* magazine that a 1984 special law on Venice meant the precautionary principle could be applied to block Project MOSE: 'In language that presages the precautionary principle, the law calls for all interventions to be "experimental, gradual and reversible".'[75]

The application of the precautionary principle to Project MOSE brought environmental precaution on to the political stage and led to further delays. This started when environmentalists in Venice's City Council demanded that the mobile gates be scrutinised by an Environmental Impact Study (EIS). The government's Comitatone Committee agreed, even though projects of this kind were not subject to Italian law on environmental impact assessment. The EIS stalled Project MOSE between 1995 and 1997. But this was only the first instance of an Italian government adopting environmental precaution for Project MOSE. Subsequently, the International College of Experts was hired to assess the EIS, delaying Project MOSE until 1998.

When Italy's government called in experts to assess an assessment, years after engineers concluded their risk assessments, political concerns with environmental precaution became detached from the safety of Project MOSE itself. The institutionalisation of the precautionary principle enabled governments to appear highly concerned about the safety of their citizens on an emotive issue, although they were held in rather low esteem in many traditional aspects of politics. The precautionary principle was employed as a political tool by Italian politicians, who were suffering a legitimacy crisis.[76]

The precautionary principle has been applied to many aspects of the Venetian lagoon. For example, the International Navigation Association Report on Environmental Management for Ports suggested a role for the precautionary principle in managing Venice's port: 'When the risks of serious or irreversible environmental damage are high and cost penalties are low, the Precautionary Principle is easily justified. In other circumstances when the risks are lower it may be better to undertake further scientific research rather than invoking the Principle.'[77]

Even perceptions of man's relationship to the lagoon have been changed by the application of the precautionary principle to Project MOSE. Giovanni Mazzacurati, as General Manager for the CVN company in charge of Project MOSE, suggested:

In the context of the Venice Lagoon, the concepts of 'fate' or 'emergency' should no longer arise, or at least they should no longer be formulated in the customary terms ... We can in fact predict the effects of hidden problems before the results become obvious and lay our hands on solutions which have been adequately studied, tested and made effective.[78]

But is this risk-aversion – manifested through Project MOSE – a sensible strategy when the risks of sinking and flooding to Venice are imminent? There are, of course, risks to implementing Project MOSE, but these risks are minimal in relation to the benefits of its construction. Classical notions of risk were based on mathematical probabilities.[79]

Unfortunately, such balanced risk assessments have been replaced by the precautionary principle, which encourages extreme caution but may indeed exacerbate safety concerns by attempting to deny that risks exist. The adoption of the precautionary principle by all the key players regarding Project MOSE – environmentalists, Venice City Council, successive governments, the CVN company, the port authorities – can only further delay its construction.

Is it 'unnatural' to be ambitious for Venice?

Venice is deserving of our efforts to protect it – it is sinking, and we should do something about it. But those with a preservationist mentality – such as the opponents of Project MOSE – believe that we can have a world without trade-offs. According to this mindset, intervening in nature for human benefit is inconceivable as it changes the state of the lagoon. Stefano Lorenzi of WWF Italy criticised Project MOSE 'for its inefficient environmental impact in maintaining the equilibrium of the lagoon's ecosystem'.[80]

They baulk at doing what is necessary for Venice's protection to sustain a mythical harmony between man and nature. But how can human societies develop if we do not separate nature from society? Italia Nostra's Venice website complains: 'Unfortunately, too many politicians, planners, architects and engineers have sought to "modernise" Venice and its Lagoon over the past century ... Italia Nostra has promoted a different vision: the projects that define the future of Venice should be designed with the water in mind.'[81]

Indeed, people throughout the entire history of Venice have sought modernisation and could hardly ignore the water. But to environmental groups, parting the seas with mobile barriers is an arrogant attempt to dominate nature that will undoubtedly change marine and plant life in the lagoon. However, transforming the lagoon is exactly what we need to do to protect Venice.

The construction and maintenance of Venice has been one of the greatest engineering achievements of mankind.[82] The unusually close relationship between a city and water-dominated conditions has required a high level of human intervention.

Conclusion

Something must be done to protect Venice from flooding. Project MOSE plus diffuse/*insulae* measures will not prevent all flooding indefinitely. But they provide the best solution for the foreseeable future. Project MOSE has been debated for over 30 years and the leading authorities had approved its design by 1994. Standard risk assessments by engineers were favourable and international experts from leading engineering companies approved it in 1993. Since 1995, environmentalists in Italian NGOs and the Green Party have led the campaign to stall Project MOSE. Through their role in governments, their initiation of the Environmental Impact Study, and court actions, they have promoted a precautionary approach towards Project MOSE.

The precautionary principle has become the modus operandi for Project MOSE under successive governments. Agreeing to the EIS and calling in the International College of Experts delayed the project from 1995 to 1998, but it then took until 2003 to definitely approve it. Even Berlusconi's current government made concessions to environmental precaution as it inaugurated the floodgates. All the key players in relation to Project MOSE have adopted the precautionary principle. This suggests that a culture of environmental precaution has superseded rational risk assessments and become an automatic response for the management of such projects. Environmental precaution has already become influential in plans to build the world's longest suspension bridge to Sicily[83] and is likely to hinder the completion of Project MOSE in 2011.

Venice has played a key role in the climate change debate, and fears over Venice 'sinking' because of rising sea levels have been blown out of proportion by the news media. This misrepresentation has been aided by widely questioned projections for total sea-level rises by the IPCC and some academics that have predicted very high increases in the relative sea level for Venice.

Given that the major problem with Venice in the twentieth century was subsidence – i.e. sinking – we should prioritise our actions to limit flooding due to sinking and rising sea levels. While there will be costs to actions, attentive management will minimise the impact of frequent gate closures on shipping and pollution in the lagoon.

How exactly the earth's climate will change is uncertain, and the consequences for humanity are also uncertain – but the solution is not to focus on how to restrict humanity's interaction with nature. If sea levels do rise, then Venice will have to adapt. Such actions are more likely to complement the protection provided by Project MOSE than to make it redundant.

Rather than fearing climate change and resorting to climate alarmism, we should employ a rational assessment of how climate changes may affect us, and then we can adapt with appropriate responses. Project MOSE is a positive, adaptation-oriented response, which will protect the historical city of Venice from being overly affected by flooding.

The sea was formerly considered a great advantage for Venetians in terms of salt, marine life, defence and as a means of transport: it enabled Venice to become the gateway to the Orient despite a constant battle to protect the city from the lagoon, rivers and the sea.[84] Now these waters are perceived as a threat and this view is perpetuated by the idea that humans cannot and should not interfere with nature. In reality, the lagoon continues to be a great advantage to the city for its tourist trade and over US$1 billion a year generated by the port. Project MOSE and the complementary measures should aid the reconstruction of a positive attitude towards the lagoon, the sea and human-inspired change.

3 Climate change: the 21st century's most urgent environmental problem or proverbial last straw?

Indur M. Goklany

Some have argued that the Kyoto Protocol and other schemes for immediately mitigating greenhouse gas (GHG) emissions are justified because human-induced global warming is, in the words of the 42nd US president, William J. Clinton, 'the overriding environmental challenge' facing the globe today.[1] Another argument, advanced by those who are more cautious and perhaps less prone to hyperbole, is that the impacts of global warming – on top of myriad other global public health and environmental threats – may prove to be the proverbial 'straw that breaks the camel's back'. They suggest that climate change will overwhelm human and natural systems by increasing the prevalence of climate-sensitive diseases, reducing agricultural productivity in developing countries, raising sea levels, and altering ecosystems, forests and biodiversity worldwide.[2]

This chapter examines whether analyses of the impacts of global warming into the foreseeable future support these arguments and, if they do, whether it is more effective to rely on mitigation strategies, or on adaptation to their impacts. In this chapter, adaptation implies measures, approaches or strategies that would help cope with, take advantage of, or reduce vulnerability to the impacts of global warming.

Global warming impacts to the present

Over the last century or more, the earth has warmed 0.4–0.8°C, perhaps due to man's influence, according to the Intergovernmental Panel on Climate Change (IPCC).[3] Over this period, there have been changes in many climate-sensitive environmental indicators or sectors of the economy – some for the better, others for the worse, and for others, neither better nor worse.

The good

For many critical climate-sensitive sectors and indicators, matters have actually improved, especially during the last half century.[4] Global agricultural productivity has never been greater, for instance. An acre of cropland sustains about twice as many people today as it did in 1900, and it sustains them better. Based on nutrition and affordability of food, people have never been fed better or more cheaply. Between 1961 and 2001, global food supplies per person increased 24%, although global population almost doubled;[5] and between 1969–71 and 1998–2000, the number of people in developing countries suffering from chronic hunger declined from 35% to 17% or, in absolute terms, from 917 million to 799 million despite a 79% growth in their population.[6]

In wealthier countries, deaths due to climate-sensitive infectious and parasitic diseases are now the exception rather than the rule. Such deaths are declining in most developing countries thanks to better nutrition and public-health measures. Accordingly, from 1960 to 2000, the global infant mortality rate dropped by 57%, and global life expectancy at birth increased from 50.2 to 66.5 years.[7] However, in the last 10–15 years, these improving trends have been reversed in many sub-Saharan African and former communist countries, not because of climate change, but because of increasing poverty, Aids and malaria.[8,9]

The bad

For other climate-sensitive indicators matters have, indeed, worsened, but so far human-caused warming has had little to do with these declines. Consider sea-level rise. Mean sea level is rising at a rate of about 0.1–0.2 mm per year.[10] While it is not known what fraction, if any, might be due to any human-caused warming, the IPCC's Science Assessment notes that there is no detectable acceleration of sea-level

Figure 1   **US property losses due to floods (1903–97)**

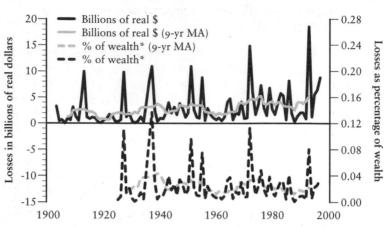

*Wealth measured as fixed reproducible tangible assets.
Source: Goklany (1998b, 2000a)

Figure 2   **US property losses due to hurricanes (1900–97)**

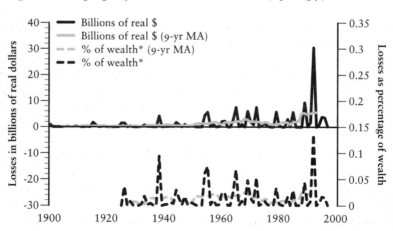

*Wealth measured as fixed reproducible tangible assets.
Source: Goklany (1998b, 2000a)

Figure 3   US death rates due to various extreme weather events
(1900–97), deaths per million population, 9-year moving
averages (MA)

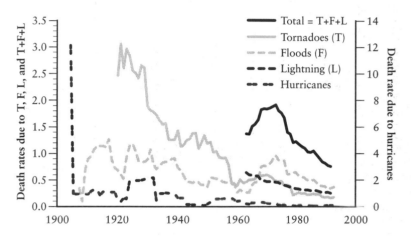

Source: Goklany (1998b, 2000a)

rise during the twentieth century.[11] Suffice it to say, so far any accelerated sea-level rise due to man-made warming is unlikely to have
caused anything other than a minor impact on human or natural systems compared to other environmental stresses such as development
of coastlines, conversion of lands for aquaculture, drainage for other
human land uses, sediment diversion due to dam construction, construction of seawalls, and subsidence owing to water, oil and gas extraction.[12]

Agricultural demand for water, probably the largest threat to freshwater species, continues to increase.[13] Meanwhile, threats to terrestrial biodiversity – primarily the conversion of habitat to agricultural
uses[14] – have not diminished. Forested area declined by 124 million
hectares (306 million acres) in tropical and subtropical nations between 1990 and 2000.[15] This decline, which occurred largely because
increases in food demand outstripped increases in agricultural yields,
is unrelated to global warming. Yet during the same period, forest
cover in the rest of the world (mainly wealthy nations) expanded by

28 million hectares (69 million acres), mainly because technology-based, high-yield agriculture has reduced the demand for cropland in those countries.

The indifferent

As the higher latitudes have become warmer, spring has arrived earlier since the 1960s. As a result, we observe earlier breeding or first singing of birds, earlier arrival of migrant birds, earlier appearance of butterflies, earlier choruses and spawning in amphibians, earlier shooting and flowering of plants.[16]

This has been accompanied by later arrival of autumn and autumn colours in some places. A meta-analysis of trends for 99 species of birds, butterflies and alpine herbs, found significant range shifts averaging 6.1 km per decade towards the poles.[17] It also found a significant mean advancement of spring events by 2.3 days per decade based on data for 172 species of shrubs, herbs, trees, birds, butterflies and amphibians.[18]

Clearly, there have been changes, but are these changes adverse? The Finnish branch of the WWF notes, for example, that:

> Thanks to the warming trend, the growing season has grown ...
> At the same time the spring migration of birds, including finches, larks, wagtails, and swifts, has begun an average of ten days earlier than before.
>
> The warmer temperatures have brought new, more southerly species of butterflies to Finland. Many existing types of butterflies have extended their habitats further north.[19]

According to the Royal Society for the Protection of Birds, some birds in the UK have also become more abundant, possibly due to milder winters.[20] Similarly, the ranges of fifteen butterfly species in the UK have expanded substantially since the 1970s, 'almost certainly' because of warming (whether or not human-induced).[21] They also appear earlier in the year and some have been able to spawn an extra generation during the summer. In addition, some moths, crickets and dragonflies have migrated into the UK.[22]

With respect to vegetation, a study of the earliest flowering dates of 385 wildflower species in the UK shows that on average they bloomed more than 4.5 days earlier in the 1990s compared to their 1954–90 average, with 16% blooming significantly earlier while 3% bloomed significantly later; one plant bloomed fully 55 days earlier.[23] Similarly, the ranges of flowering plants and mosses seem to have expanded in the parts of Antarctica that have warmed.[24] Soil invertebrates have also advanced with changes in vegetation.[25]

Obviously, warming (whether due to man's activities or nature's machinations) seems to have a measurable impact on the distribution and abundance of species, but it is far from clear whether these changes are beneficial or detrimental. More importantly, the major current threats to species come from habitat modification and loss, water diversions, and invasive species, perhaps in that order.

Summary

Despite any warming, by virtually any climate-sensitive measure of human well-being, human welfare has improved over the last century.[26] While some credit for increasing agricultural and forest productivity is probably due to higher carbon dioxide concentrations and higher wintertime temperatures,[27] most of these improvements are due to technological progress driven by market- and science-based economic growth, technology, and trade. Such progress has also reduced human vulnerability to the effects of climate change.[28] As a result, technological progress has so far had a greater impact on the climate-sensitive sectors than has climate change itself.[29]

On the other hand, matters may actually have deteriorated for some climate-sensitive environmental indicators, such as the loss of habitat and forests, and threats to biodiversity. However, so far, climate change (human-induced or not), while contributing to change, does not seem to be responsible for most, if any, of this deterioration.

Therefore, it is difficult to sustain on the basis of current evidence the notion that climate change is the greatest threat to public health or the environment today. But what about the future?

The future with and without global warming

Table 1 on page 63 allows us to assess the importance of global warming, relative to other factors that might affect public health and the environment into the 'foreseeable future'. This table is based, for the most part, on a set of impact studies sponsored by the UK Department of Environment, Food and Rural Affairs (DEFRA), many of which

have been incorporated into the IPCC's 2001 Third Assessment Report (TAR). Because the DEFRA-sponsored assessments did not provide an estimate of the future forest cover in the absence of climate change, it was necessary to rely on other studies reported by the IPCC for that estimate.

Notably, analysts involved in the DEFRA studies recognise that socioeconomic projections are 'not credible' beyond the 2080s,[30] hence the selection of the 2080s in the table as the outside limit to the 'foreseeable future'. Although the TAR states that between 1990 and 2100, global temperature might increase from 1.4 to 5.8°C, it also notes that 'on time scales of a few decades, the current observed rate of warming ... suggests that anthropogenic warming is likely to lie in the range of 0.1 to 0.2 °C per decade over the next few decades.'[31]

However, the scenarios employed in the DEFRA-sponsored impact assessments are based on globally averaged temperature increases of slightly more than 0.3°C per decade;[32] therefore, they may overestimate likely impacts to the 2080s.

In Table 1, column 3 provides estimates of various public health and environmental risks or factors related to those risks under baseline conditions in the 2080s (i.e., in the absence of global warming). Column 4 provides the changes in risks or risk-related factors in the 2080s due to the imposition of global warming, above and beyond baseline conditions; that is, it provides estimates of the reductions in total risks if climate doesn't change after 1990. Finally, column 5 provides estimates of reductions in total risks or risk-related factors due to full implementation of the Kyoto Protocol assuming that because the Protocol would reduce temperature change between 3–7% by 2100,[33] it would reduce the impacts of global warming by less than 7% for all risk categories except coastal flooding. For the latter, it is assumed that the Protocol will decrease the impact of climate change by thrice that amount.[34]

Table 1 indicates that:

- In the absence of warming, (that is, in the 'baseline' case), *global cereal production* would increase by 123% from 1,800 megatons in 1990 to 4,012 megatons in the 2080s in order to meet additional food demand of a larger and wealthier global population.[35] Such an increase is plausible provided agricultural technology continues to enhance productivity,

Table 1 Projected climate change impacts in the 2080s, compared to other environmental and public health problems

Climate-sensitive sector/indicator	Year	Impact/effect — Baseline (B) in the 2080s, includes impacts of environmental problems other than climate change	Impacts of climate change (ΔCC) in the 2080s, on top of the baseline	Impacts of Kyoto Protocol in 2080s, relative to baseline+ –ΔCC*
Global agricultural (cereal) production	2080s	4,012 million metric tons (MMT), vs 1,800 MMT in 1990	production would drop 2% to 4%; and could be substantially redistributed from developing to developed countries	increase net global production by 0.1% to 0.3%
Falciparum malaria (population at risk, PAR)	2080s	8.82 billion at risk by the 2080s, vs 4.41 billion in 1990	increase PAR by 0.26 to 0.32 billion (or 2.9% to 3.7%)	reduce total PAR by 0.2% to 0.3%
Water resources (population in countries where available resources use > 20%)	2085	6.46 billion, vs 1.75 billion in 1990	increase PAR from 0.04 to 0.11 billion (or 0.6% to 1.6%)	reduce total PAR by about 0.1% or less
Global forest area	2050s / 2080s	decrease 25–30(+)% by 2050, relative to 1990	increase by 5%, relative to 1990	reduce the increase in global forest area by 0.4%
Sea-level rise (SLR)	2080s	varies	~40–41 cm (or 20 in), relative to 1990	reduce SLR by <1.4 in
Coastal flooding (PAR)	2080s	0.013 billion	increase PAR by 0.081 billion (or 623%)	reduce total PAR by 18.1%
Coastal wetlands (area)	2080s	decline of 40% relative to 1990	decline of 12% relative to 1990	reduce the decline by 0.8%, relative to 1990 level
Storms	2080s	unknown	unknown whether magnitudes or frequencies of occurrence will increase or decrease in any specific area	unknown

Sources: Parry *et al.* (1999) and IPCC (2001b) for agriculture; Arnell *et al.* (2002), and Goklany (2000), based on Solomon *et al.* (1996), for forest cover; Arnel *et al.* (2002) and IPCC (2001b) for *Falciparum* malaria; Arnell *et al.* (2002) for coastal flooding; Arnell (1999) for water resources.
* Assumes that the Kyoto Protocol, if implemented, would reduce climate change and its impacts by 7% by the late 21st century. See text.

sufficient investments are made in the agricultural sector and related infrastructure, and trade continues to move food from surplus areas to deficit areas.[36]

Due to global warming, agricultural production may decline in poor countries, but may increase in wealthy countries, resulting in a net decline in global production of 100 megatons to 160 megatons (i.e., 2–4% of total production in the absence of warming) in 2080. Thus, downturns in economic growth (which would inhibit investments in the agriculture and infrastructure), slower technological change, or less voluntary trade of food supplies are more likely to create a future food crisis than any potential global warming.[37] Notably, the Kyoto Protocol would result in a marginal improvement in production of less than 0.3% in the 2080s.

- The *population at risk (PAR) of malaria*, one of the most common and dreaded climate-sensitive infectious diseases, might essentially double in the absence of global warming, from 4.41 billion in 1990 to 8.82 billion in 2080.[38] With global warming, the numbers at risk of contracting malaria might increase by 0.26 billion to 0.32 billion in the 2080s (equivalent to an increase of between 2.9% and 3.7% over the 2080 baseline).[39] However, an increase in the numbers at risk does not necessarily translate into increased number of cases of malaria, or its prevalence.[40] The Kyoto Protocol would reduce the total numbers at risk of contracting malaria by less than 0.3% in the 2080s, as well. (For a more detailed explanation of climate and disease, see Chapter 1, 'Could global warming bring mosquito-borne disease to Europe?')

- The number of people living in countries who experience *water stress* (defined as countries using more than 20% of their available water resources) would increase from 1.750 billion in 1990 to 6.464 billion under the baseline (no-climate-change) case in the 2080s.[41] The latter number would increase by between 0.042 billion and 0.105 billion, depending on the precise climate model employed to estimate future climate change.[42] The impact of the Kyoto Protocol for this risk category will also be minimal into the 2080s.

- If all else remains the same (i.e., ignoring changes in land use after 1990), then by the 2080s, *global forest area* may *increase*

5% over 1990 levels due to global warming alone.[43] But if greater agricultural and other human needs increase the demand on land, as they well might (since the world's population will be larger and probably wealthier), forest cover may decline by 25–30%, putting enormous pressure on global biodiversity.[44]

- *Sea level* may rise about 40 cm from 1990 to 2080.[45] As a result, the population at risk of *coastal flooding* is expected to increase by 623% from 0.013 billion under the baseline to 0.094 billion. The Kyoto Protocol could reduce the total PAR from coastal flooding by about 18%. Sea-level rise could also lead to a loss of *coastal wetlands*, but such losses due to other human activities are expected to dominate at least into the 2080s.

- It is unclear whether the frequencies and magnitudes of *storms*, such as tornadoes, hurricanes and cyclones, will increase or decrease.[46]

Thus, with the exception of coastal flooding, the impacts of climate change into the foreseeable future are secondary to the impacts of other agents of change built into the baseline case. Moreover, for the most part, the impacts of global warming would seem to be within the noise level of these baseline problems.

Consequently, stabilising GHG concentrations immediately – even if feasible – would, unfortunately, do little over the next several decades to solve the bulk of the problems frequently invoked to justify actions to reduce humanity's role in warming. Land and water conversion will continue almost unabated, with little or no reduction in the threats to forests, biodiversity, and carbon stores and sinks. Food production would not be markedly increased or decreased. Populations at risk of malaria would not be affected much, nor would the numbers of people at risk of water stress. The reductions in risks due to the Kyoto Protocol would be relatively trivial, at least until the 2080s, with respect to all risk categories – again with the exception of coastal flooding.

Nevertheless, climate change could be the proverbial last straw. Moreover, the relatively large reductions in the PAR from coastal flooding might arguably, by itself or in conjunction with reductions in other risk categories, justify the Kyoto Protocol (or other mitigation schemes).

Dealing with the last straw

There are several approaches to solving the problem of the straw that might break the camel's back; none of them needs to be mutually exclusive.[47]

Focusing on the last straw

The most common approach is to focus almost exclusively on the last straw, especially on reducing or eliminating it. This is equivalent to reducing or eliminating climate change, i.e., by reducing or eliminating GHG emissions. However, as Table 1 shows, this would accomplish little except in the case of coastal flooding, because it would leave untouched the major share of the total risk burden.

Reducing the cumulative burden

Another approach would be to lighten the total burden on the camel's back before it breaks. This is tantamount to reducing the cumulative environmental burden before global warming causes significant and irreparable damage. Consider malaria, for instance. Under the first approach, mitigation, one would, at most, eliminate the 0.26–0.32 billion increase in the PAR from malaria in the 2080s by eliminating climate change – which is impossible. By contrast, under the second approach, one would attempt to reduce the total PAR from malaria, whether it was 4.41 billion in 1990 or 9.14 billion in the 2080s. This approach has several advantages.

First, even a small reduction in the baseline (or non-climate-change-related) PAR could provide greater aggregate public-health benefits than would a large reduction in the relatively minor increase in PAR due to climate change. Assuming that annual cases and deaths due to malaria vary with the PAR, reducing the base rate for malaria by an additional 0.3% per year between 1990 and 2085 would compensate for any increases due to climate change.

Second, resources employed to reduce the base rate would provide substantial benefits to humanity decades before any significant benefits are realised from limiting climate change. Considering that 1 million Africans currently die from malaria every year, and that its death toll can be cut in half at a cost of between US$0.38 billion to US$1.25 billion,[48] humanity would be better served if such sums were spent now to reduce malaria in the near future, rather than on limiting climate change only to curb a relatively minor share of the potential increase in malaria decades from now. Moreover, the benefits of reducing malaria in Africa today with the second approach are real and far more certain, and Africans would experience these benefits decades sooner than any benefits resulting from eliminating climate change.

Third, the technologies developed and public-health measures implemented to reduce the base rate would themselves serve to limit additional cases due to climate change when, and if, they occur.

Fourth, reducing the baseline rate would serve as an insurance policy against adverse impacts of climate change, whether that change is due to anthropogenic or natural causes or if the changes occur more rapidly than currently projected. In effect, by reducing the baseline today, one would also help solve the cumulative malaria problem of tomorrow, regardless of its cause.

Fifth, because of the inertia of the climate system, it is unrealistic to think that future climate change could be completely eliminated. Moreover, the Kyoto Protocol experience indicates that because of its socioeconomic impacts, even a freeze in emissions is likewise unrealistic, despite the inability of such a freeze to actually halt further climate change.

The logic of reducing the cumulative burden applies to other climate-sensitive problems and sectors where factors unrelated to climate change are expected to dominate for the next several decades. As Table 1 indicates, these problems and sectors include agriculture, food security, water, forests, ecosystems, and biodiversity.

Increasing resilience and reducing vulnerability

Yet another approach to dealing with the last straw is to strengthen the camel's ability to carry a heavier load. This calls for improving resilience and reducing vulnerability.

It is generally acknowledged that poorer countries have the greatest vulnerability to climate change, not because their climates are expected to change the most, but because they lack the resources to adapt adequately to any change. But their expected difficulty of coping with climate change is only one manifestation of a deeper overarching problem, namely poverty.

If we look around at the world today, we find that almost every indicator of human or environmental well-being improves with wealth (see Figure 4).[49] This is true whether or not the indicator is climate-sensitive. Poorer countries have less food available per capita; they are

Figure 4 **Human well-being vs economic development, 1990s**

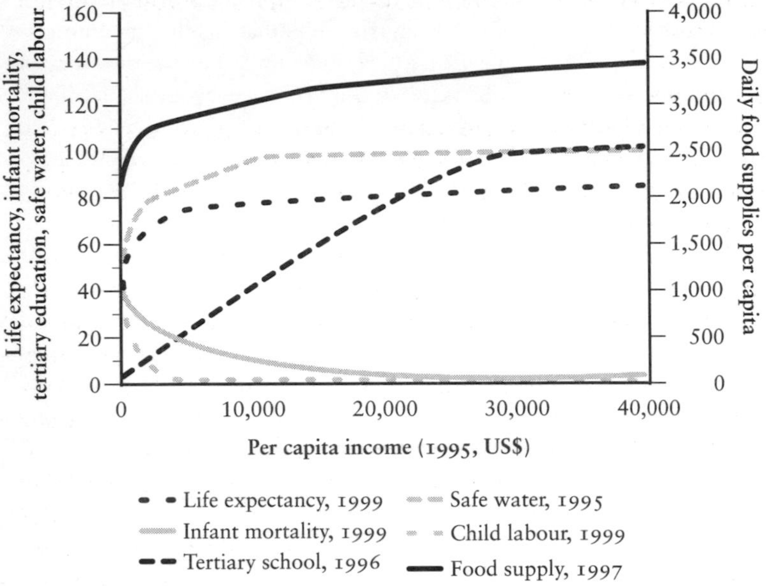

Source: Goklany, 'Affluence, Technology and Well-being', *Case Western Reserve Law Review* 53: 369–390 (2002)

hungrier and more malnourished; their air and water are more polluted; they have poorer access to education, sanitation and safe water; and they are more prone to death and disease from malaria and other infectious and parasitic diseases. Consequently, they have higher mortality rates and lower life expectancies.[50]

These populations are more vulnerable to any adversity because they are short on the fiscal and human-capital resources needed to create, acquire and use new and existing technologies to cope with that adversity. As a consequence, economic growth, by enhancing technological change, would make society more resilient and less vulnerable to adversity in general, and to climate change in particular.

Focusing on enhancing economic growth should be complemented by efforts to bolster the institutions that underpin society's ability and desire to develop, improve and utilise newer and cleaner technologies.

ions includ⸱   viding greater protections for property
                                                        the rule of law, providing honest and
                                                        ⸱ureaucracies, and supporting freer and

About International Policy Network

International Policy Network (www.policynetwork.net) is a UK-based
charity which coordinates policy activities to broaden public understanding of issues relating to sustainable development, health, technology and
trade. To achieve that goal, IPN sponsors events and publications, coordinates activities at international meetings, and promotes greater media
awareness of public policy issues.

IPN's Sustainable Development Project promotes the view that real
sustainable development is about promoting progress, eliminating
poverty, and empowering people through the institutions of a free society,
because such policies allow people and communities to take charge of
their own lives.

IPN is also a capacity-building organisation which works with numerous organisations and individuals in developing countries to enable them
to participate more effectively in public policy activities.

egional winners and losers. In particu-
⸱cultural production, with developing
developed countries producing more.
⸱lem of hunger in the former.
⸱duction are not a new problem. Nor is
⸱ly, poor countries consume about 10%
[2] Their future dependence on food im-
⸱heir demand for food is expected to
⸱ral productivity. Such imbalances have
⸱nd large, through trade. Freer trade
⸱uture whether the imbalance is caused
⸱ctor. In effect, trade is akin to helping
⸱traw by sharing the burden amongst
⸱ping countries would need the where-
⸱es produced elsewhere. This is yet an-
⸱conomic growth, particularly in the
⸱ries.[53]

to Annex B countries in 2010 is esti-
⸱% of their GDP.[54] Let us assume that
⸱% of GDP. In 2000, that would have
⸱5 US dollars.[55] Due to these expendi-
⸱zero or less (see below), would rise
⸱enefits would include a reduction of
⸱ooding, but for the other climate-sen-
⸱le 1, the Protocol's benefits would be

⸱lising GHG concentrations would be
⸱s. Stabilisation at 450 ppm (parts per
⸱ted to cost a few trillion dollars be-
⸱, despite the considerable costs, and

regardless of which mitigation regime is imposed, the formidable baseline problems indicated in Table 1 would be virtually undiminished for all risk categories except coastal flooding. Moreover, one should expect that some residual impacts of global warming would persist.

Notably, some studies suggest that temperature increases of the order of 1–2°C might, in fact, result in net benefits for agricultural and timber production.[56] Consistent with these assessments, the IPCC report also suggests that a small (~1–2°C) increase might possibly be a net economic benefit to the world but an increase in excess of 2°C could be negative.[57] Therefore, the costs of any mitigation may have to be shouldered for several decades before one can be confident that they would create net benefits. This problem is magnified because of the inertia of the climate system, which magnifies the lag time between when emission reductions are initiated, and when a noticeable effect on the impacts of global warming is observed.

On the other hand, instead of mitigation-based approaches, we could employ a set of adaptation strategies based upon the principles outlined above for dealing with the last straw, and targeted to each of the risk categories in Table 1. These strategies would enhance adaptability and/or reduce vulnerability to both the impacts of warming and the other global changes included in the baseline. They would also have the added advantage that the benefits would be observed sooner after the costs were experienced. Examples follow.

- The global cost estimate for protecting against a 50 cm *sea-level rise* in 2100 is about US$1 billion per year,[58] or less than 0.005% of the overall global economic product.[59] Compared to the Protocol and any other mitigation approach, this is orders of magnitude cheaper, and would also provide greater reductions in the PAR from *coastal flooding* into the 2080s and beyond.
- A 20% increase, for example, in global agricultural research funding, which in the mid-1990s stood at US$33 billion per year (including US$12 billion in developing countries)[60] ought to, over 95 years, more than erase the entire 4% reduction in *agricultural production* due to global warming (this would be substantially more than the trivial portion that the Protocol would restore).

Figure 5   Net habitat loss to cropland vs. increase in agricultural
             productivity, 1997 to 2050

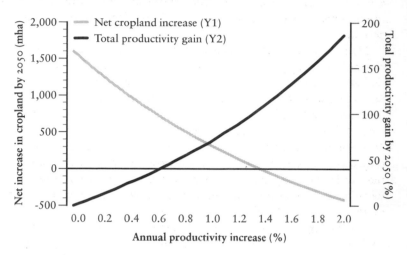

Source: FAO (2000) per Goklany (1998a, 1999a)

- No less important, to the extent the additional research
  funding increases sustainable agricultural productivity beyond
  the quantity needed to replace the shortfall, it would reduce
  the human demand for land. Figure 5 shows that increasing
  agricultural productivity would not only reduce conversion of
  wild land to new cropland, but it could return existing
  cropland back to nature. Increasing agricultural productivity is
  the single most effective method of preventing habitat loss and
  fragmentation, and conserving *global forests*, terrestrial
  biodiversity and *carbon stocks and sinks*.[61]
- Similarly, the above increases in agricultural research, if
  targeted appropriately, would also help to increase the amount
  of food that can be produced by one unit of water. Since
  agriculture is responsible for 85% of the fresh water consumed
  globally, each 1% reduction in agricultural water consumption
  allows consumption for other sectors to increase by 5.7%.
  This would not only reduce the PAR from *water stress* but also
  would decrease pressures on *freshwater species*.

- Annual expenditures of between US$0.38 billion and US$1.25 billion could reduce the current death toll from *malaria*, about 1 million people per year according to the World Health Organization. This too would be far more effective in reducing death and disease from malaria than either full implementation of the Kyoto Protocol or even halting climate change altogether. As previously noted, methods developed to prevent or treat baseline malaria problems (that is, the problems in the absence of global warming) can be used to address similar problems resulting from climate change.

Thus, until the 2080s, the above set of adaptation measures would cumulatively cost much less, and deliver greater benefits, than either the Kyoto Protocol or other more ambitious mitigation schemes.

Advocates of immediate GHG controls, however, might argue that regardless of the urgency of climate change during the next several decades, unless GHG emission reductions commence now those reductions may come too late to do any good. The reason, they claim, is the inertia of the climate and energy systems.

But as Table 1 indicates, even if 50 years are required to replace human energy systems from start to finish,[62] we could nevertheless wait an additional 25 years or more before initiating control actions to produce change beyond that which would be obtained automatically through continuous, long-term improvements in technologies. Meanwhile, we could implement the strategies outlined above, which would deliver benefits for people living today, while enhancing our ability to address future problems that climate change may exacerbate or cause. These strategies could be complemented by developing more cost-effective mitigation and adaptation technologies that could be implemented when they are needed.

Conclusion

Assessments of present-day and future impacts of human-induced climate change indicate that it is not now, nor is it likely to be in the foreseeable future (i.e., into the 2080s), as significant as other environmental and public health problems facing the globe. Nevertheless, global warming could be the proverbial straw that breaks the camel's back, particularly for natural ecosystems and biodiversity.

But instead of merely focusing on lightening or eliminating the last straw – analogous to reducing or halting climate change – the camel's back may also be saved in other ways. Also, reducing or eliminating the last straw does little good if the camel's back bends or breaks in the meantime.

Instead of concentrating on the last straw to reduce the cumulative burden of possible problems, we should reduce today's urgent public health and environmental threats (such as malaria, water stress, hunger and habitat loss) that might be exacerbated by climate change. As we have seen, this would provide greater, more cost-effective and quicker benefits to both humanity and the rest of nature.

We should also strengthen the camel's back so that it can withstand a heavier load, regardless of how or why the load was generated. The basic reasons as to why some societies are less resilient and more vulnerable to climate change are precisely the same reasons why they are also less resilient and more vulnerable to adversity in general, namely, they have insufficient economic development and a lower propensity towards technological change.

Not surprisingly, poorer countries with less ability to develop, afford and use new technologies have higher rates of hunger; poorer public health services; greater incidence of infectious and parasitic diseases; less access to education, safe water or sanitation; and, therefore, greater mortality rates and lower life expectancies. Accordingly, we should strengthen the institutions that drive both economic growth and technological change. Not coincidentally, many of these institutions nurture and foster each other. This approach would enhance societies' abilities to cope not only with climate change, but adversity in general, regardless of its cause, or whether it is man-made or not.

Thirdly, we should make it possible to share the burden among numerous camels. Since climate change would create regional winners and losers, the burden could be spread more evenly through trade. Thus shortfalls in agricultural production induced by climate change in some countries could be addressed through trade with others that would experience gains in agricultural production. Trade, moreover, has the added benefits of stimulating both economic growth and technological change. It particular, it allows societies everywhere to gain from innovations and inventions made elsewhere in the world, without having to reinvent wheels.

Policies based on these alternative approaches, all of which rely on improving adaptability and reducing vulnerability, are superior to the single-minded pursuit of reductions in climate change, at least into the foreseeable future. Into the 2080s, they would provide greater benefits, far sooner and far more economically than would be achieved by efforts which focus on mitigation.

Indeed, by reducing vulnerability and increasing adaptability, we might – consistent with the stated ultimate goal of the UN Framework Convention on Climate Change – raise the level at which GHG concentrations might need to be stabilised to avoid dangers to humanity and nature, which would further reduce the costs of addressing climate change.[63]

Despite the inertia inherent in both the climate and energy systems, we have at least two to three decades before we need to embark on socially and economically costly efforts to reduce GHG emissions that would go beyond 'no-regrets' actions.[64] In the interim, we should focus on: (a) solving today's urgent problems while creating the means to address future potential problems due to climate change; (b) improving our understanding of the impacts of climate change so that we can distinguish between the possible and the probable; (c) increasing information regarding the trade-offs and synergies between adaptation and mitigation; (d) reducing barriers to implementing no-regret technologies, whether they are related to mitigation or adaptation (such as eliminating needless subsidies for energy and natural resource uses); and (e) undertaking efforts to expand the portfolio of no-regret actions through greater R&D into more cost-effective mitigation and adaptation technologies.

Such a multifaceted and holistic approach would help solve today's problems to improve the lives of people living today, without compromising our ability to address future challenges, whether caused by human-induced climate change, another agent of global change, or something else entirely.

Part Two

Strategies for Adapting to Climate Change

4 Is Kyoto a good idea?

Martin Ågerup

Introduction

In 1997 the Kyoto Protocol ('Kyoto') was signed by 172 countries. It is by far the most ambitious international environmental treaty ever attempted, with the purpose of controlling emissions of greenhouse gases (GHGs) related to human activities, mainly carbon dioxide (CO_2) from carbon-based fuels. The regime intended to reduce emissions from industrialised countries so that during the period 2008–12, they are 5.2% beneath their levels in 1990.

Although there was agreement in principle to this regime, since 1997 the path has been extremely rocky. Though the USA decided in 2001 not to ratify the treaty, the other signatories have decided to forge ahead. Most of the participants, including the countries of the European Union, have now ratified the treaty, which will come into force if Russia ratifies in the autumn of 2003.

The Kyoto Protocol resulted from the concern that if humans exacerbate the greenhouse effect, the earth's climate and human living conditions will suffer. The intended benefit of the treaty is to prevent the potential negative effects of a warmer climate, such as extreme weather events and sea-level rise.

Kyoto attempts to achieve this goal by limiting human emissions of greenhouse gases in industrialised countries, which is currently provided by the burning of hydrocarbon-based fuels and therefore results in considerable emissions of GHGs. Such a policy does not come without costs to society – our economic production, prosperity and lifestyle depend on access to cheap and reliable energy. Therefore the agreement has significant costs for our everyday lives.

While Kyoto may be based on good intentions, good intentions do

not make sound public policy. What matters for Kyoto is expected results. We do not know what will happen in the future, but it is worth
asking 'Does the Kyoto Protocol appear to be a good idea given reasonable expectation of the future based on the best available current
knowledge?'

The purpose of this chapter is to answer this important question,
with an analysis of the expected costs and benefits of the Kyoto Protocol, amidst significant uncertainties and unresolved scientific issues.
To achieve this, we ask: How significant are man-made emissions of
greenhouse gases? Does the expected climate change pose a problem?
How will Kyoto address this problem, and at what cost? Fundamentally, is the Kyoto Protocol the right approach to address climate
change?

How significant is man-made warming?

Scientists are in broad agreement that human activities have some influence on global mean temperatures and climate, but they disagree
about the extent of this influence. Even without the influence of humanity, the earth's climate would not be stable – it experiences extreme natural changes, and small man-made temperature increases are
likely not to be a huge problem.

It is almost as uncontroversial that uncontrolled human emissions
of GHGs will result in a rise in global mean temperatures – all other
things being equal. This does not in itself justify political action in
general or the Kyoto Protocol in particular. The earth's climate would
not be stable without human influence. In fact climate has always
changed and will always change. Therefore, small man-made temperature increases are not a problem.

To evaluate the Kyoto Protocol, it is important to establish by how
much temperatures can be expected to rise over a relevant timescale
because of the influence of human activities. The Intergovernmental
Panel on Climate Change (IPCC) has attempted this exercise for the
past fifteen years. The IPCC was established in 1988 by the World
Meteorological Organization (WMO) and the United Nations Environment Programme (UNEP). The role of the IPCC is to assess:

> The scientific, technical and socio-economic information relevant
> to understanding the scientific basis of risk of human-induced

climate change, its potential impacts and options for adaptation and mitigation. The IPCC does not carry out research nor does it monitor climate related data or other relevant parameters. It bases its assessment mainly on peer reviewed and published scientific/technical literature.[1]

In its third assessment report (2001), the IPCC predicted that the global mean surface temperature would rise by 1.4–5.8°C[2] and came to the widely quoted conclusion that: 'There is new and strong evidence that most of the warming observed over the last 50 years is attributable to human activities.'[3]

The IPCC's conclusions are highly controversial. Its prediction depends on two sets of computer models, one of which predicts how much CO_2 and other greenhouse gases human activities will emit into the atmosphere, and the other of which predicts how the climate will react to such increases.

The quality of the predictions depends entirely on the ability of the models to produce reliable results – but in fact, there are two main uncertainties which hamper these models. We do not know how sensitive the earth's climate system is to increased levels of CO_2, nor do we know how much CO_2 we will emit.

How sensitive is the earth's climate system to CO_2?

It may seem surprising that this should be a controversial question since the basis of greenhouse theory rests on solid scientific knowledge. It is controversial because the expected effect of increased levels of CO_2 on climate can be divided into direct and indirect effects. The direct effect is a function of the radiative properties of CO_2 (its ability to trap infrared light), which are fairly well established. But this direct effect of increased CO_2 is relatively small. The IPCC estimates the direct effect of a doubling of CO_2 in the atmosphere would be on the order of 1.2°C with an accuracy of +/-10%.[4] The major part of the CO_2-warming predicted by computer models is based on indirect feedback effects.

The most important indirect feedback effects are caused by water vapour. According to the IPCC: 'In models, increases in water vapour ... are the most important reason for large responses to increased greenhouse gases.'[5]

Water vapour is by far the most important GHG. Humans 'emit' water vapour, but in minute amounts compared to natural evaporation from oceans and the earth's biomass. This evaporation is not a problem, since water does not accumulate in the atmosphere like other GHGs. Unlike CO_2, methane and other GHGs, there is a limit to how much water vapour the atmosphere can carry – that is why water vapour turns into rain, fog or dew and carbon dioxide does not. But when temperatures rise, the atmosphere absorbs more water vapour, so an increase in CO_2 leads to a small rise in the global mean temperature. Higher temperatures increase humidity, and since water vapour in the atmosphere is itself a greenhouse gas there is a further rise in temperature. That is a feedback effect.

The problem for climate modellers is that, unlike the direct radiative effect of CO_2, indirect feedback effects are not well understood. To estimate the indirect effects, we need a detailed understanding of the climate system itself, but we lack that understanding because the climate system is extremely complex.

We know that there are both positive and negative feedbacks in the system. Water vapour forms clouds. Some types of cloud have a net warming effect, and others have a net cooling effect. Some clouds trap reflected infrared heat which is 'escaping' the atmosphere, so they have a warming effect. Others reflect incoming solar radiation back into space and thus have a cooling effect. To model the feedback effect of clouds, modellers need to know how clouds are formed and how they behave – but they don't. The IPCC admits in its latest assessment report that 'Clouds represent a significant source of potential error in climate simulations ... The sign of the net cloud feedback is still a matter of uncertainty ...'[6] and

> Probably the greatest uncertainty in future projections of climate arises from clouds and their interactions with radiation ...
> increased physical veracity has not reduced the uncertainty attached to cloud feedbacks: even the sign of this feedback remains unknown.[7]

Thus, a huge degree of uncertainty surrounds climate modelling, which implies that the major part of the warming predicted by climate models is uncertain.

In fact, recent research indicates that the positive feedback effects from clouds may be vastly exaggerated in current models. A few years ago, Professor Richard Lindzen, meteorologist at the Massachusetts Institute of Technology, published with colleagues from NASA a peer-reviewed article which highlights the extent of uncertainty about clouds.[8]

The researchers found that the area of cirrus clouds over the equatorial Pacific Ocean correlated negatively with temperature. In other words, the production of cirrus clouds went down as temperature went up. The net effect of cirrus clouds is to trap heat, so if the statistical correlation represents a climate mechanism, Lindzen may have found a powerful negative feedback which current climate models do not account for. It would be so powerful that it would reduce the effect of a doubling of CO_2 from the current IPCC estimate of 1.5–4.5 °C to 'a half degree or maybe one point-something', according to Lindzen.[9]

Most of the warming predicted by climate models results from built-in assumptions that positive feedbacks will considerably outweigh negative feedbacks. These assumptions are also uncertain, and they are contradicted by research conducted by Lindzen and other scientists.

Indeed, chapter 7 of the latest IPCC report deals extensively with the physical processes underlying the presumed feedback effects and points out many problems with the way computer models represent these processes. But in the 'Summary for Policymakers' this 35-page chapter is summarised in one sentence under the heading 'Confidence in the ability of models to project future climate has increased'. That sentence reads: 'Understanding of climate processes and their incorporation in climate models have improved, including water vapour, sea dynamics and ocean heat transport.'[10] While this is undoubtedly true, it is not the whole truth. Some quotes from the executive summary of chapter 7 put that sentence into context:

> While improved parametrizations have built confidence in some areas, recognition of the complexity in other areas has not indicated an overall reduction or shift in the current range of uncertainty of model response to changes in atmospheric composition.

Probably the greatest uncertainty in future projections of climate arises from clouds and their interactions with radiation ...

... increased physical veracity has not reduced the uncertainty attached to cloud feedbacks: *even the sign of this feedback remains unknown.* [emphasis added]

... significant deficiencies in ocean models remain.

Uncertainty resides with the relative importance of feedbacks associated with processes influencing changes in high-latitude sea surface temperatures and salinities, such as atmosphere-ocean heat and fresh water fluxes, formation and transport of sea ice, continental runoff and the large-scale transports in ocean and atmosphere.

There has been an increase in uncertainty in those aspects of climate change that critically depend on regional changes.[11]

How much CO_2 does humanity emit?

The IPCC emissions scenarios
The climate models calculate the consequences of increasing atmospheric GHG concentration – typically at a level equivalent to doubling of atmospheric CO_2 compared to the pre-industrial level. However, to forecast expected warming, modellers need estimates of the rate of increase of CO_2 to determine when CO_2 concentrations will double, which means that they need forecasts of human emissions of CO_2 throughout the 21st century.

The IPCC has carried out such long-range forecasting exercises for all of their three assessment reports since 1992. The latest results were published in 2000 in the Special Report on Emissions Scenarios (SRES).[12] The SRES working group produced 40 scenarios, out of which six were chosen as 'marker scenarios' and fed into the climate models. The emissions scenario exercise is a crucial step in creating the above-mentioned temperature growth range for 2100 of 1.3–5.8°C. In fact, most of the span in this temperature range is produced by the huge difference in emissions across the marker scenarios.

The scenarios are based on parameters which influence emissions, such as GDP per capita growth; population growth; energy efficiency (how much GDP does one unit of energy produce); composition of fossil fuel consumption (coal, oil, gas); and non-carbon fuels' share of total energy production (nuclear, wind, solar, etc.).

Use and abuse of scenarios

The quality of the emissions forecasts depends entirely on how these parameters are treated and their values in the different scenarios. Over a century, the degree of uncertainty for each parameter is enormous, and these uncertainties compound. However, this does not mean that anything should be considered plausible.

A variety of problems such as sloppiness and poor methodology plague the use of economic scenario methods in many applications, and the IPCC is no different. Unfortunately the Special Report on Emissions Scenarios, part of the IPCC, does a very poor job of creating plausible scenarios.

The IPCC treats the period 1990–2000 as part of the forecast period, without consideration for the actual data that exist for that period. Apparently the SRES reused data from previous scenario exercises, which have been proven wrong by time and real world observations.

For instance the scenario figures used for the increase in world GDP between 1990 and 2000 vary between 20.6% and 35.4%. However, IMF data for most of that period was available in 1999 and showed growth of 36.5%.[13] This error probably does not have a substantial impact on model results by 2100, but it is amazing that such sloppy practices are used as an input to a modelling exercise that involves the use of supercomputers and costs millions of euros.

The IPCC also suffers from poor methodology. One leading economic modeller, John Reilly of the MIT Joint Program on the Science and Policy of Global Change, calls the SRES approach to scenario building an 'insult to science'.[14] According to Reilly the scenario teams have worked backwards from a desired end result in terms of emissions and temperature increases. In other words, the IPCC has allegedly started with an emission projection, then made an estimate of the relationship between emissions and growth, and finally calculated the growth rate needed to achieve the desired emissions projection.[15]

Economists Ian Castles and David Henderson recently called attention to a specific deficiency in methodology, observing that the method for calculating future GDP was based on an inappropriate assumption about exchange rates between countries. This seemingly innocuous methodological sleight of hand led to an overestimate in future emissions.

Castles and Henderson concluded that: 'The SRES projections do not, as is claimed for them, encompass the full range of uncertainties about the future ... The SRES should not be taken as the accepted basis for the IPCC's coming Fourth Assessment Review.'[16]

The most serious objections to the SRES result from their use of scenarios as forecasts. The SRES states that they are 'images of the future or alternative futures' and should not be seen as predictions or forecasts, but then it feeds the emissions scenarios into a computer and produces an apparently concrete result about what the temperature will be. Though the SRES presents these scenarios as an exercise in 'free thinking' about the future, the IPCC relies on them as direct evidence for the need to drastically curtail global CO_2 emissions.

The SRES claims that the scenarios 'are not assigned probabilities of occurrence, neither must they be interpreted as policy recommendations'.[17] But since each scenario results in a figure, there is an implicit bias in favour of extreme scenarios, because they extend the temperature range.

The SRES scenarios are used to make forecasts – so the SRES team abuses the scenario method. In the real world, extreme outcomes are treated as less likely and therefore are assigned less importance. The SRES scenarios do not have assigned probabilities. This is a normal practice when working with scenarios, since such exercises are intended to cover a full range of possible futures. Yet it implies that scenarios should not be used as forecasts, because a forecast doesn't make sense without a discussion of probability. Meteorologists only make weather forecasts extending three to five days into the future. The reason why they don't make longer forecasts is that there is only a very small probability that they will be correct.

How much warmer will the climate be by 2100?
Table 2 summarises the basic steps that the IPCC goes through to achieve the expected warming interval (left-hand column). For each

---

Table 2 IPCC's forecasting and uncertainties

Forecast exercise	Principal uncertainties
Create scenarios for future emissions of CO_2	World GDP growth per capita and its distribution among low-, middle- and high-income countries
	Population growth
	Composition of different fossil fuels in total consumption
	Technological change, including shifts to non-carbon or low-carbon energy sources, energy efficiency, carbon sinks etc.
Convert emissions to atmospheric concentrations	The lifetime of different GHGs in the atmosphere
Model radiative forcing and convert this forcing to a projected temperature	The sensitivity of the climate system to increased CO_2 (feedback effects of clouds, aerosols etc.)
	Natural climate effects enhancing or counterbalancing man-made effect

---

step it lists widely considered uncertainties given current scientific understanding (right-hand column).

We simply do not know how much warmer the climate will be in 2100. In fact, the degree of (compound) uncertainty is so large that the mere exercise undertaken by the IPCC of providing temperature intervals is highly misleading and provides phoney confidence. For many of the parameters even the degree of uncertainty is controversial. Climate science is not yet capable of providing confidence intervals, especially one hundred years into the future. In fact even the term 'uncertain' is often misleading when it comes to climate science – many things are not uncertain but simply *unknown*. For all climate scientists know, climate might have cooled by the year 2100!

The National Academy of Sciences report concludes:

Because there is considerable uncertainty in current understanding of how the climate system varies naturally and reacts to emissions of greenhouse gases and aerosols, current estimates of the magnitude of future warming should be regarded as tentative and subject to future adjustments (either upward or downward).[18]

The climate sensitivity of the models is largely based on a number of positive feedbacks, whereas negative feedbacks are probably underestimated or completely ignored. The temperature range should be decreased because climate models almost certainly overestimate the sensitivity of the climate system to increased CO_2. On top of that, the emissions scenarios produce growth rates of carbon emissions which are not in line with recent history. Given this knowledge, downward adjustments seem to be the best option, but there are strong indications that the IPCC has a systematic bias in favour of exaggerated temperature increases.

Will warming be a problem for humans?

Advocates of the Kyoto Protocol often justify this policy on the basis that climate change will adversely affect human beings. So, from a human welfare perspective, what might be the effects of climate change?

Using a human-centred approach also implies using a dynamic approach. With the right incentives and institutional framework, people solve problems, they rise to challenges and adapt. They invent new technologies and new ways of doing things. And over the course of an entire century they will have plenty of time to do so. Some of the climate change models assume that people don't adapt to new conditions – that is not a reasonable assumption.

A rise in sea levels

The IPCC expects sea levels to rise 9–88 cm by 2100 – a wide-ranging figure, and, given scientific uncertainties, the high figure should be taken with a pinch of salt. Even if sea levels were to rise by as much as a metre, it would be a minor problem. In some parts of the world, people would have to adapt to avoid flooding. The IPCC predicts that damage to infrastructure in coastal areas would costs tens of billions of dollars. However, these estimates are based on the rather absurd assumption that people and countries will just idly sit by and watch. Instead countries would act to protect coastal areas – just as the people of Holland have done for centuries. This adaptation can be done with relatively low costs. The IPCC estimates that for most countries the costs of protection are in the region of 0.1% of GDP.[19]

The US Environmental Protection Agency has estimated that the cost of coastal protection against a whopping 1-metre rise in sea level would be just 1.5% of one year's GDP at current levels of output.[20] Since sea levels rise gradually, the investment could be spread out over 50 to 100 years. On top of that, most of the investment would take place in a future where the US GDP would be anywhere from five to ten times higher than the present. In other words, even a 1-metre sea level rise would have little impact on human welfare, even if developed countries were to finance all adaptive measures in developing countries.

Extreme weather events
This is one of the most popular misconceptions about the effects of climate change. Despite much media spin, there is no empirical evidence that hurricanes and storms are increasing in frequency or intensity.[21] The US has fewer hurricanes during hot El Niño years than during normal years.[22] Climate models do not predict these developments: '... climate models currently lack the spatial detail required to make confident projections. For example, very small-scale phenomena, such as thunderstorms, tornadoes, hail and lightning, are not simulated in climate models.'[23]

Although global warming is expected to cause increased rainfall in many parts of the world, increased problems with flooding are almost entirely linked to developments such as urbanisation, sewage systems leading rain water quickly into rivers, conversion of wetlands, deforestation etc.

The higher insurance costs of weather-related natural disasters are, according to insurance company Munich Re, caused not by global warming but by a growing world population. More and more people can afford insurance, and they have more property to lose and increasingly live in higher-risk coastal areas.[24]

Changes in agricultural output
If climate changes without farmers adapting their behaviour accordingly, global warming would be detrimental to agricultural output.

But it is not reasonable to assume that farmers would not adapt to changing conditions. If reasonable assumptions about adaptive behaviour concerning crops and technologies are included, global warming is expected to benefit agricultural output in industrialised countries, because a warmer and more humid atmosphere is good for plant growth and because CO_2 enhances the growth of plants.[25]

According to models used by the IPCC, the benefit in agricultural output will mainly be in developed countries in the cooler northern hemisphere. In the Third World, the models expect a negative impact because of the rise in (already high) temperatures. IPCC assumptions about adaptation seem far too conservative: 'These studies do not fully or adequately account for technological progress and adaptation ... which not only overestimates the potential negative impact but also *underestimates the potential gains from harnessing positive aspects of global warming*'[26] (emphasis added).

The IPCC uses a completely unrealistic worst case scenario,[27] with a 4.0–5.2°C temperature increase as early as 2060, in which only current technologies and crop varieties can be used, and no new technologies – crops, irrigation methods, pesticides or farm machinery – will be invented. But even this nightmare scenario produces an outcome where developing countries only lose 6–7% of agricultural output.

It seems likely that with adaptation and technological advances, less developed countries will not suffer a loss in agricultural output because of global warming. In fact, the models probably overestimate the negative effect from rising temperatures, because most of the actual warming is occurring at night.[28] This will extend the growing season, while the risk of drought and other heat wave damage is reduced.

Consequences for human health

Global warming is not likely to have a negative effect on human health. Humans have successfully adapted to varying climates. There is no general temperature level at which heat suddenly becomes dangerous to human health. On the contrary, heat-related mortality increases when the temperature rises above what the local population is accustomed to. In Finland heat-related mortality sets in at 17.3°C, in Athens at 25.7°C.[29]

Over the course of a century, humans will adapt to rising temperatures, or they will adapt their environment to the temperature, and they will suffer no adverse health effects. In fact, since death rates due to extreme cold are double those due to extreme heat, there might be a net benefit from warming in the number of lives saved.[30]

Some alarmists promote the idea that tropical diseases will spread because of global warming. However, the geographical spread of these diseases has very little to do with climate. In the nineteenth century, malaria, cholera and other diarrhoeal and parasitic diseases were prevalent around the world, including northern Europe. Malaria was endemic in England until the late 1800s and in Finland until after World War II. Today these diseases are problems 'only in countries where the necessary public health measures are unaffordable or have been compromised.'[31]

Wealth and a functioning public health system is what matters when it comes to combating tropical diseases.

The costs and benefits of the Kyoto Protocol

The Kyoto Protocol was adopted with the objective of establishing a regulatory framework that could tackle the threat of global warming. The main mechanism for achieving this is quantified emissions commitments. Only industrialised countries with high per capita emissions, listed in Annex 1 to the Treaty, are bound by the emissions agreement.

The Annex 1 countries include the USA, OECD countries in Europe, Japan, Canada, Australia, New Zealand and economies in transition: Russia, Ukraine, Poland and Hungary, amongst others.

These countries agreed to cut their emissions of GHGs. Each country agreed to achieve a specific percentage reduction by 2008–12 relative to 1990 (the base year) emissions. The USA agreed to cut emissions by 7% and the EU by 8%, and Japan agreed to a 6% reduction. The EU reduction was later divided internally among the EU countries. Some countries were actually allowed to increase emissions (especially poor southern European countries) while others took deeper cuts to compensate.

The negligible benefits of Kyoto

The benefit of the Kyoto commitments comes from the extent to which emissions reductions avoid the costs of global warming. Additionally there might be tangential benefits, for instance reduced air pollution. The costs of the Kyoto Protocol are the costs of limiting energy use. The expected benefits are virtually all in the long run, since global warming would happen gradually as GHGs accumulate in the atmosphere, but the costs are paid immediately.

The main expected benefit of Kyoto-type initiatives is that less CO_2 causes less warming. However, the size of this benefit is extremely uncertain because we do not know how much warming we will avoid (i.e. the benefit) from incurring the costs of Kyoto, nor do we know if the warming we are avoiding is actually detrimental. If Kyoto avoids the marginal effects of a small amount of warming (as seems almost certain – see below), it would probably not be beneficial at all, even if it could be implemented at no cost.

Given current (lack of) knowledge it does not even make sense to calculate the benefits of Kyoto: they are not just uncertain, they are unknown. They could be positive or negative. No one knows, and we cannot know until we achieve a better understanding of both climate and climate change.

The costs of Kyoto in the EU

It is an established fact that the Kyoto commitments will have little impact on emissions and global warming. There are two reasons for this. First, the agreed reductions amount to just 5.2% reduction below the 1990 level for the industrialised countries overall. This will not stop the accumulation of GHGs in the atmosphere. Second, and more significantly, although currently Annex 1 countries are responsible for most of global GHG emissions, this situation will change as populous Third World countries such as China and India grow richer. Over the next decades increasing emissions from less developed countries will completely drown out the small reductions achieved by Kyoto.

This has caused some critics of Kyoto to ridicule it as a futile exercise. But that misses the point: the Kyoto Protocol is far more than just a specific commitment to reduce emissions by a certain date. It provides a global mechanism for *future* commitments, and commits parties, for instance, to open negotiations on a second commitment period no later than 2005. It is the general understanding of government officials and academics participating in or following the Kyoto process that the Protocol is a mechanism which will progressively seek deeper cuts in carbon emissions, and that commitments will eventually expand geographically to include all countries.

There are two relevant periods for estimating the costs of the Kyoto Protocol – the first commitment period, until 2012, and then beyond 2012. While it is possible to estimate the costs of Kyoto during the first commitment period, those of a post-Kyoto regime are highly uncertain.

The first commitment period, 2008–12

The costs of the carbon reduction commitments depend on the extent to which the parties to the treaty are allowed to trade CO_2. Trading CO_2 implies that countries are allowed to sell or buy rights to emit CO_2. A market for CO_2 will evolve and CO_2 will be traded at a market price. Some economic agents will be able to cut CO_2 emissions cheaply – and they can make money by selling their extra emission rights on the market. Others, who face higher costs of cutting emissions, would happily buy these rights. This reduces the overall costs of curbing emissions, because the market allows for CO_2 emissions to be reduced where it costs the least.

In October–November 2001 the seventh Conference of the Parties in Marrakech agreed on a regime that allows emissions trading[32] and a carbon trading market is now evolving.[33] The consequences for total abatement costs are significant: several studies show that a regime allowing trade among all Annex 1 countries would reduce abatement costs between 50% and 75%.[34]

Trading is the most cost efficient way for EU countries to fulfil their Kyoto obligations – and the ultimate costs to them depend on whether or not they will use trading. During most of the negotiations, the EU countries were against trading, as were most environmental organisations. They feared that trading would lead to countries buying 'hot air' in Russia and eastern Europe instead of actually making an effort at home to save energy and develop more efficient technologies.

The 'hot air' issue arises because Russia and Eastern Europe have experienced dramatic economic collapses just after the 1990 base year. This resulted in decreases in energy consumption and emissions. Despite the fact that these economies have grown considerably over the past 6–8 years, emissions have not returned to previous levels.

Emissions in these countries have actually fallen far below their Kyoto commitment, which is a reduction of 0–8% (depending on country) relative to the base year (1990), so these countries have excess carbon to sell on the market. They may need this carbon in the future, but that does not seem likely (at least in the short term) given recent emission trends in these countries.

The price of carbon depends on the supply and demand. In 2001 three events drastically changed the supply and demand for carbon. First, the USA – by far the largest source of potential demand for carbon – withdrew from ratifying Kyoto. Second, Russia's energy projection was revised downward thus increasing their carbon surplus. Third, the Bonn/Marrakech negotiations introduced a CO_2 allowance of 'carbon sinks' into the Kyoto regime, which allows countries to subtract carbon absorbed by managed forests from their emissions. Russia's sink allowance was almost doubled, which further increased its surplus carbon.

The result of these developments was that the price of carbon plummeted. Recent price estimates range between US$5 to US$13 per tonne.[35] Assuming that European countries decide to fulfil their Kyoto obligations by buying carbon, and that the carbon allowance in Russia and eastern Europe is sufficient and traded, compliance with the Kyoto Protocol until 2012 would be very cheap, because actual emissions reductions will be minute. Kyoto 2008–12 is estimated to reduce global CO_2 emissions by less than one per cent relative to no policy.[36]

Such a reduction would have no measurable effect on climate whatsoever. Virtually all the emissions reduction has already taken place in Russia and eastern Europe during the transition from central planning to market economy, and the costs of this reduction have also been carried by these economies. The costs of Kyoto will not be incurred by efforts to cut emissions but will be a transfer of money from Europe (and Japan) to Russia and eastern Europe.

The EU as a whole would have to spend between €940 million and €2.6 billion if they were to comply with Kyoto today. In nominal terms that is a lot of money, but it is only 0.01–0.03% of EU GDP. However, it is a yearly cost, and it would increase as economic activity – and thus emissions 'demand' – increases.

This would be the cheapest possible scenario, but several factors are likely to increase the costs considerably.

It is uncertain that the complete allowance from Russia and eastern Europe will be available for trading, and economic models show that the future trading price is very uncertain. One study shows that increasing the assumed GDP growth rate in Russia by 1% for ten years would cause the carbon price to increase by 50%.[37] The fact that just one variable has such a large influence on carbon prices shows how vulnerable the carbon market will be. Also, eastern Europe and Russia could decide to bank their CO_2 allowance for use in the future. This could happen if the price becomes too low or if they collude to push up prices. In the latter case Kyoto compliance would become more expensive.

Carbon prices are vulnerable, but given the supply and demand situation, it is difficult to imagine carbon trading costs multiplying.

What could increase costs is if the EU countries decide to finance a major share of emissions cuts on their own. While the Kyoto Protocol does allow trading, article 6 (d) provides the following limitation on trade: 'The acquisition of emission reduction units shall be supplemental to domestic actions for the purposes of meeting commitments under Article 3.'

This provision is vague, which leaves it open to interpretation. In practice some countries are interpreting it strictly, possibly for ideological or political reasons. Holland and Italy have announced that they will fulfil only half their targets through trade.[38] In Denmark the government decided in early 2003 to use the trading regime to minimise costs of abatement.[39] However, this decision was ferociously criticised by the left wing opposition. At the time of writing legislation has been passed by the European Parliament which will allow an EU-wide trading scheme to go into effect in January 2005.

If the EU or individual countries choose not to utilise the trading regime, costs will increase significantly. This is because national measures would actually reduce CO_2 emissions rather than simply requiring businesses to buy 'hot air'.

A large number of economic models have estimated the costs of Kyoto under different assumptions. One of these assumptions[40] is a regime where no trade is allowed. This is similar to a situation where countries decide to curb CO_2 by using national measures despite the existence of a trade regime. The lowest estimate indicates a loss of 0.31% of GDP per year in OECD Europe. The highest estimate shows a 2.08% loss.[41] For OECD Europe this would be equivalent of €34–230 billion a year.

The probable outcome is that the 2008–12 commitment period will cost much less than expected before Marrakech and the US opt-out – probably less than half a percentage point of GDP. However, the effectiveness of the money spent in terms of carbon reduction will be very low. One of the leading economic modellers suggests that 'the overall assessment of the accord is that it pays a high price for very small reductions in carbon emissions.'[42]

Beyond 2012

Extending the current Kyoto regime into the future would be far more expensive in subsequent commitment periods than in the initial 2008–12 period. There are two main reasons for this. First of all, the large allowance of CO_2 in Russia and eastern Europe will gradually disappear as their economies grow and emissions rise. This will result in a steep increase in carbon prices. The expected low carbon price of the first commitment period is a result of unique circumstances – including the US withdrawal – which are unlikely to recur.

Secondly, the marginal costs of CO_2 abatement increase rapidly. The costs of cutting the first tonne of CO_2 are low but rapidly it becomes more difficult to find cheap reduction options. Denmark is a case in point. Its GHG reduction efforts have been relatively intense for about a decade, and its CO_2 emissions have nearly stabilised. A recent study concluded that: 'Denmark has ... been focussing on CO_2 emissions reductions for some time now. As a result of this the measures that are possible today and lead to real reductions are relatively expensive.'[43]

Even if Kyoto is merely extended into the future, stabilising Annex 1 country emissions at 5.2% below the 1990 level, costs will increase exponentially. The reason for this is that, as economies grow, new ways of avoiding GHG emissions will have to be found in order to stabilise emissions, which is costly. Alternatively – and even more costly – economic growth would have to be avoided.

One study used a computer model to analyse a scenario where the Kyoto Protocol is extended forever (for the whole of the 21st century) in the Annex 1 countries.[44] The Protocol would permanently decrease European GDP through the 21st century. After ten years GDP would be approximately 0.1% below the 'no Kyoto' reference case. After twenty years the gap would be 0.35% and the gap would continue increasing for the first 50 years and reach a maximum of slightly over 0.5%.

Fractions of a per cent may not sound like a lot, but the actual amounts involved are astronomical. The study estimates net global costs of the Kyoto Protocol to Annex 1 countries to be between US$344 billion and US$1,507 billion in current value.[45] In comparison, global Third World aid amounts to approximately US$60 billion.

As mentioned above the IPCC estimate of the global costs of warming will be in the order of US$500–US$650 billion a year. Compared to this figure the Kyoto investment might seem a good idea. However, Kyoto is estimated to have a very small impact on climate and temperatures. Assuming IPCC level climate sensitivities would imply that Kyoto had the effect of lowering temperature by 0.15°C.

The 'Kyoto forever' scenario would, despite the high costs, have very little impact on total emissions. If we were to stabilise emissions or temperature, the cost would be much higher.

Given the mechanisms of the Kyoto Protocol and the current political agenda, it is possible that future commitment periods will imply deeper cuts in GHG emissions in order to stabilise CO_2 emissions or global temperatures at some future level by some future date. The costs of cutting GHG emissions beyond 2012 depends on the quantity, the methods for cutting them, and how fast.

Conclusion: How should we address climate change?

Dealing with climate change implies dealing with uncertainty. After all, we are discussing the effect of current actions and policies on society and climate almost a hundred years from now.

In fact, this paper concludes that the benefits of curbing emissions of GHGs are not just uncertain but *unknown* since we do not know the extent to which GHGs influence climate. Also, a small temperature increase is probably good whereas a large temperature increase is probably bad. Without any clear benefits, Kyoto could actually be bad for human welfare, even if it had no costs.

Current scientific knowledge suggests that the CO_2 sensitivity of the earth's climate is probably lower than suggested by computer models – at least the high-end models. This supports the possibility that by curbing emissions we might in fact be preventing something good. On the other hand, we cannot rule out the possibility that increasing GHG concentrations will result in large and detrimental temperature increases.

Turning from benefits to costs, estimates of future costs of climate mitigation rely on assumptions about future economic and population growth, and about future technologies, which have implications for the carbon intensity of energy use and the energy intensity of the economy

Beyond 2020 the assumptions about these factors are even more uncertain, especially for technology. The main technological uncertainty is not *if* low-carbon or non-carbon energy technologies will become economically competitive, but *when*.

McKibbin and Wilcoxen, two leading economic modellers and both professors of economics, point out 'the importance of the inherent uncertainty about the future that should be at the heart of the design of a sustainable climate policy.'[46] They criticise Kyoto's rigid emissions targets on a fixed timetable, which results in highly uncertain costs for countries. As mentioned above, the authors demonstrate how changing just one factor (the assumed GDP growth rate in Russia) would cause the carbon price to increase by 50%.

Over the long run costs could explode if expected new technologies were delayed by a few years. Countries would have to comply with the treaty obligations, though the rational course might be to postpone emission reductions until new technologies are in place. Such a policy would free economic resources for further cuts, or for other uses.

According to McKibbin and Wilcoxen, imposing rigid targets and timetables implicitly assumes 'that the risks posed by climate change are so great that emissions must be reduced no matter what the costs. However, too little is known about the dangers posed by climate change, and about the costs of avoiding it, to draw that conclusion.'[47] Given the uncertainties about the costs of both climate change and abatement strategies, we need to balance the risk of doing too much with the risk of doing too little. Current scientific knowledge about climate change does not warrant the Kyoto-style approach of 'reductions at any cost'.

But on the other hand GHG concentrations may possibly entail some adverse effects. Since climate change is (presumably) irreversible, it makes sense to avoid causing more of it than necessary, at least until the potential risks are better understood.

An appropriate policy response to climate change would be one that encourages all low-cost efforts to slow GHG emissions. The policy should provide an incentive to reduce GHG emissions but should avoid imposing unknown or very large costs. A strong economy is important for dealing with any kind of change, and costly policies can create a drag on the economy. Moreover, a strong economy will provide the means for research into less carbon-intensive technologies. The main reason why Third World countries are more vulnerable to global warming than industrial nations is because they are poorer and their societies sometimes do not encourage appropriate responses to change. Thus, poor countries are less able to control their environment.

Others have observed that Kyoto's rigid approach is much less efficient than alternative regimes. A study by William Nordhaus, one of the leading modellers of the economic effects of Kyoto, concludes that: 'The emissions strategy is highly cost-ineffective, with the global temperature reduction achieved at a cost almost 8 times the cost of a strategy which is cost-effective in terms of "where" and "when" efficiency.'[48]

The 'where' efficiency of Kyoto is low because only Annex 1 countries participate, despite the fact that most of future emissions – and the cheapest abatement potential – will come from Third World countries. The 'when' efficiency is low because rigid targets and timetables impose 'too much abatement, too soon'. The IPCC favours early action: 'The greater the reductions in emissions and the earlier they are introduced, the smaller and slower the projected warming and rise in sea levels.'[49]

The IPCC failed to add in its advice to policy makers that the greater and earlier the reductions, the higher the costs.

In the long run the key to addressing climate change is technology. We need to develop new low-carbon or non-carbon energy technologies and we need to increase the energy efficiency of our economies. It is almost certain that we can develop cheap non-carbon energy alternatives by the second half of the 21st century. But in order to do this we need prosperous economies that can afford to invest in research and development. We also need to minimise uncertainty.

The Kyoto Protocol is the wrong approach to addressing climate change. While it does provide an incentive for developing new decarbonising technologies, it is highly inefficient, because its rigid targets and timetables demand cuts at any cost. Furthermore, Kyoto actually maximises uncertainty – the future price of carbon, future commitments and the number of participating countries are all unknown factors. This regime of uncertainty is not conducive to creating a long-term strategy for future energy technologies.

5 Sustainable energy for the poor

Barun S. Mitra

Introduction

While people in wealthy countries are obsessing about consuming too much energy, several billion poor people do not consume enough energy. Energy poverty – characterised by a lack of affordable, reliable and plentiful energy sources – affects over one-third of the world's population, mostly poor people in rural areas (where 70% of the world's poor people live).

These people depend on traditional forms of energy that are of low intensity and cause harm to both the environment and human health. There are huge opportunity costs which have not been given enough consideration in the climate change debate – particularly because these are urgent problems today rather than hypothetical long-run problems that might be caused by global warming.

But some environmental groups have argued that poor countries should not take the same developmental path as wealthy countries, because it could cause pollution. Greenpeace and The Body Shop, in a campaign for the 2002 World Summit on Sustainable Development in Johannesburg, proclaimed that 'Oil, coal and gas cannot meet the needs of the poorest, but "positive" or renewable energy can.'[1] Elsewhere, one author claims that 'Economic globalization is accelerating [global warming] by accelerating industrial activity and universalizing the carbon-intensive model of development worldwide.'[2]

Of course, these statements are simply patronising: to live more sustainably, poor people need to consume more energy and more resources. There is a direct negative causal relationship between lack of energy, the burden of poverty and environmental degradation in rural areas.

Energy and rural development are mutually dependent, and they represent one aspect of the poverty cycle that pervades most rural areas in India. Breaking this deadlock is one of the major challenges that developing countries face in developing their rural areas. It is likely that problems resulting from lack of energy will only be alleviated by investment in facilities that provide energy on a wide-scale basis. India is already full of small-scale informal sector entrepreneurs who are providing energy to the poor, because poor people are willing to pay for it.

This chapter focuses on the economic, environment and health impacts of dependence on traditional energy in India. In particular, it argues that poor people need to consume more energy to improve their lifestyle and health, to add value to their economic activities and to eliminate poverty. Though it may seem counterintuitive, increased energy consumption induces higher energy efficiency, and also leads to better environmental protection.

However, the chapter does not advocate any single type of energy as a panacea. Whether small or large, whether decentralised or grid-based, small generators, hydropower or hydrocarbons, biomass or wind – anything which is produced by the energy market should be allowed to flourish in India to solve the energy needs of the poor.

Energy scenario: global and Indian

Global

Present trends in energy consumption, and the projected future, show a steep increase in energy demand. Global primary energy demand is projected to increase on an average by 1.7% per year from 2000 to 2030, reaching an annual level of 15.3 billion tonnes of oil equivalent from the current level of 9.1 billion tonnes.[3] The outlook further states that the share of developing countries in total energy demand will increase from the current level of 30–43% while that of the developed countries will fall from 58% to 47%.[4] The share of traditional energy in industrialised countries is below 2% of the total, while reaching 46% in South Asia and 53% in sub-Saharan Africa.[5]

High-income group countries consume 51% of the total commercial energy consumed in the world and account for 80% of the income generated. Middle-income group countries consume 36% of the total energy while generating only 17% of the total wealth. Low-income

Table 3	Trends in India's energy consumption (Petajoules)[8]				
	1970	1980	1990	2010 (projected)	2020 (projected)
Coal	1070	1444	1827	3704	6444
Petroleum	649	1212	2218	7604	15551
Natural Gas	12	33	283	742	950
Hydroelectricity	175	323	762	3124	5965
Nuclear	8	11	22	-----	-----
Fuel wood	2345	2864	3210	5722	6723
Crop residues and dung	1286	1568	1764	3101	3474
Total	5545	7455	10086	23457	39107
Total in MTOE*	132	177	246	558	930

*Metric tonnes of oil equivalent

group countries consume 13% of the total energy and generate only 3% of the total wealth.

Heavy reliance on the traditional energy has been one of the factors for slow economic growth in the developing countries and this factor has also contributed to low consumption and a low per capita output of commercial energy in the developing countries.

India

In 2000, India's total energy demand stood at 1.5 million tera joules (MTJ). About 65% of that was generated by commercial sources, and the rest came from non-commercial sources such as fuel wood and agriculture wastes. Households are the major consumers with nearly 44% of total energy consumption, followed by industry at around 40%.

India's per capita energy consumption is relatively low. In 1999, per capita energy consumption in India was 12.3 million Btu, compared to 355.8 million Btu (British thermal units) per person in the United States and a world average of 63.6 million Btu per person.[6]

According to one study which analysed India's 1991 National Census: '95% of the rural population still relied primarily on biomass fuels (dung, crop residues, and wood). A small fraction uses coal, which means about 97% of rural households relied principally on

Table 4 Sectoral and per capita use of biomass energy in India[11]

Types of traditional energy	Domestic sector (mt/year)		Industry & estabs (mt/year)		Total use (mt/year)	Per capita (GJ/year)
	Rural	Urban	Rural	Urban		
Fuel wood	252	30	6	10	298	5.29
Crop residue	99	–	–	57	156	2.68
Dung cake	109	5	–	–	114	1.85

these unprocessed solid fuels. Nationwide, some 81% of all households relied on these fuels; 3% used coal and 78% used biomass.[7]

Table 3 opposite shows clearly that traditional energy, i.e. fuel wood, crop residues and dung, has a dominant share in India's energy consumption.

Traditional energy

Traditional energy, which consists of biomass (wood, agricultural residues and dung), is a major energy source for about 2.4 billion people, mostly in developing countries, and particularly for rural people. Some forms of energy, such as LPG, have a broad market share in urban markets, but in rural areas only 1.3% of households use it for cooking.[9]

Traditional energy forms are usually burned inefficiently, with an open hearth or three-stone fire, which typically transfers only 5–15% of the fuel's energy into the cooking pot, and the excess is produced as smoke.[10] Generally traditional energy is used in poorly ventilated dwelling places, such as huts.

Due to a lack of commercial energy, traditional energy meets the energy demands of rural Indian households, accounting for nearly 90% of household energy use. It is generally used for cooking, while inefficient devices fuelled by kerosene are used for rural lighting. People in urban areas, in contrast, utilise commercial energy since it is available. Table 4 above compares the use of traditional energy in urban and rural areas of India.

Table 5   **Level of indoor pollutants by fuel type during cooking hours in the kitchen**

Types of fuel	Total suspended particulate (mg/m3)	Carbon monoxide (mg/m3)
Cattle Dung	2.75	144
Wood	1.98	156
Coal	1.10	94
Kerosene	0.46	108
LPG	0.46	14

Source: Ravindranath (2000), p. 38.

The health burden of India's traditional energy

Traditional energy forms have many adverse health effects, particularly for women and young children, who are disproportionately exposed to the by-products of biomass. According to one assessment, 'indoor exposures to the combustion products of unprocessed solid fuels have been estimated to produce the majority of (non-smoking) human exposures to particulates and probably to a range of other pollutants as well.'[12]

Women and children are most exposed to smoke because they spend the most time indoors, cooking and tending the fire. About 700 million women and children worldwide suffer as a result of this form of pollution.

According to a study by the World Health Organization, 'the public health problem of indoor air pollution is severe, accounting for nearly two million deaths and 53 million "disability adjusted life years" [DALYs] lost, which represents about 4.3% of the global total of DALYs lost in developing countries.'[13]

According to the WHO, 'Acute lower respiratory infections (ALRI) remain the single most important cause of death globally in children under 5 years and account for around 2 million deaths annually in this age group.'[14] This is estimated to be 6–7% of the global disease burden.[15] These deaths are largely caused by indoor air pollution, resulting from burning wood and dung. Likewise, indoor air pollution is a primary cause of morbidity.

For India, indoor air pollution constitutes '4–6% of the Indian national burden of disease, [which places] indoor air pollution as a major risk factor in the country.'[16] For women and children – 44% of the population – 'it is equivalent to 6.3–9.2% of the burden [of disease]', about two-thirds of the country's disease burden.[17]

Biomass smoke contains many thousands of potentially harmful substances. Of these, the most damaging are particulates – carbon monoxide, nitrous oxides, sulphur oxides, formaldehyde, and polycyclic organic matter, which includes carcinogens such as benzo[a]pyrene. Small particles of diameter less than 10 microns (PM10), and in particular those less than 2.5 microns (PM2.5), penetrate deep into the lungs and have the greatest effect on health.[18]

Frequent exposure to particulate matter and these chemical compounds is associated with chronic bronchitis, respiratory infections and diseases, congestive heart failure and early onset of *cor pulmonale*.[19] It is also associated with tuberculosis, adverse pregnancy outcomes, chronic obstructive lung disease and several types of cancer.

Studies have revealed that women in Indian rural areas were exposed to total suspended particulates of about 7000 microgrammes per cubic metre in each cooking period, whereas the annual standard for outdoor air is 140 microgrammes per cubic metre. The exposure to benzo[a]pyrene was equivalent to smoking ten packets of cigarettes per day. Their exposure to toxic tiny particulates during a cooking cycle is 33 times greater than that of standard ambient air pollution.[20]

Exposure to smoke during pregnancy and infancy, either through tobacco or particulate matter from biomass smoke, may therefore increase the risk of chronic obstructive pulmonary disease. One consistent finding in patients exposed to biomass is substantial carbon deposits in the lung tissue (anthracosis).[21] The adverse effects of indoor air pollution as a whole have an even worse impact on the lives of women and children, as illustrated by Table 6 on page 104.

Acute massive exposure to wood smoke can be rapidly lethal. Besides asphyxia and carbon monoxide intoxication, severe damage to the respiratory epithelium with airway and pulmonary oedema can result.[22] Another common problem found among people exposed to smoke for long periods of time is minor irritation of the eyes, which can lead to conjunctivitis.

Table 6 **Estimated annual premature deaths from indoor air pollution in India**

Category	Causes	No. of deaths annually
I.	Acute respiratory infections (age less than 5 years) Chronic obstructive pulmonary disease	310,000–470,000
II.	Blindness (women; no death) and perinatal effects Tuberculosis (women)	50,000–130,000
III.	Cardiovascular disease (women); asthma	50,000–190,000
	Grand total:	410,000–790,000*

*Total annual deaths in India for women and children under 5 in these disease categories in the early 1990s.
Source: Parikh, J. *et al.* (1999), *Economic and Political Weekly*, vol. XXXIV, no. 9, Feb–March 1999.

Economic and social cost of traditional energy

Though the direct cost of traditional energy is negligible, since it is not traded in the market, the economic and social cost of burning biomass fuels is immense. Mostly these costs come in the form of opportunity costs which are not easily quantified by economic statistics – poor health, lost time and human effort expended.

Women and young children spend hours each day in the drudgery of collecting firewood or collecting, drying, and storing manure for use in cooking, heat or light. Making dung cakes can take up to two hours a day, depending on how much dung a woman has access to and the amount of cooking fuel required.[23] Because children are involved in acquiring energy, this means they spend less time, or no time, in school. Most homes in rural villages are not connected to an electrical grid and remain dark at night, so productive evening activities, either economic or leisure, are not enjoyed.

The time that rural women spend collecting wood and performing other household tasks (which are also largely based on manual labour) leaves little time for productive employment, education and community involvement. Male household members often move to cities and towns for economic opportunities, which leads to increased numbers of female-headed households, and additional time and labour burdens for women and children in rural areas.

If women had to spend less time acquiring energy, they would probably use the time they save to care for their families, to engage in other economic activities, and to help themselves develop. This would contribute positively towards the development of rural communities.

Women's opportunity costs

Women in India spend up to 20% of their time every day collecting fuel wood.[24] Even at minimum wages, it has been estimated that the human effort spent to collect fuel wood is equivalent to 2% of India's gross national product.[25] At the national level it is found that the average number of hours spent on gathering biomass (including fuel wood, crop residues and dung) is about two hours per day, per household.[26]

At a minimum daily wage rate of Rs50 (US$1) for an adult labourer, this means that 7.5 man days (based on an eight-hour work day), equivalent to Rs375 (US$8), are spent every month collecting fuel wood. Annually, this loss amounts to about Rs4,500 (US$97).

Even if we concede that labourers will be unable to find full employment throughout the year, this is a huge income loss – about 25% of India's per capita income. The productivity gains caused by improved energy access could thus be immense.

Environmental costs of traditional energy

Traditional energies actually cause great devastation to the environment, especially in areas which are not owned by anyone, or are owned by government. Local people have little incentive to maintain these areas, and are likely to exploit the resources on those lands in an unsustainable manner. Combined with grazing and fodder-collection, all of these contribute heavily to deforestation, erosion and loss of biodiversity. And as we have already seen, these fuels contribute heavily to local environmental problems in the form of indoor air pollution. In urban areas especially they contribute to outdoor air pollution.

People's impact on forest and soils are also a key factor, with almost 25% of annual emissions of carbon dioxide resulting from forest clearance.[27]

Uprooting shrubs and trees leads to loosening of soil and becomes one of the primary reasons for soil erosion and loss of soil fertility. It also enhances the flow of rainwater during the rainy season that sometimes leads to flooding in downstream land. Massive deforestation in rural areas also has resulted in disruption of the socioeconomic life of tribal peoples. Excessive removal of vegetation and damage to ground vegetation during removal of fuel wood could affect plant diversity.

Currently, the right to collect fuel wood is one of the biggest sources of conflict between people in and around protected areas, and their managers. While these problems are the result of myriad causes (poorly defined property rights, corruption and aggressive government intervention), reducing the reliance of poor people on fuel wood would inevitably minimise conflicts between people and protected areas.

Environmental opportunity costs

The environmental costs of traditional energy should also be viewed in terms of alternative uses for resources currently used for energy. For instance, wood used as fuel may be more highly valued in for other uses, i.e. furniture, paper or lumber. Moreover, the time required for tree regeneration and forest succession means that wood is probably more valuable if it is not used as biomass energy.

Of course, it is illegal in India for private companies to fell timber for sale, which means that rural people cannot accurately assess the trade-offs between using timber as fuel, and using it for other purposes.

Lack of energy also relates to the amount of time that rural people devote to cultivation of agricultural crops; 77% of the required energy in agriculture is derived from animal and human energy. The remainder, 23%, of commercial energy consists of chemical fertilisers (14%), electricity (6%) and diesel (3%).[28]

It would be more efficient if cow dung currently used for cooking and heating were used primarily as a means to replace soil nutrients such as nitrogen. One estimate suggests that the annual nitrogen contribution from a cow or a buffalo with a mean yield of 5 kg/day would be 5.5 kg of nitrogen per animal every year.[29]

If biomass were to be used for major energy production, many millions of acres of vegetation and trees would need to be cleared. It is estimated that a 1MW grid-connected biomass combustion power plant operating 500 hours per year would require nearly 600 tonnes of dry wood (1.3 kg dry wood per kwh). At a productivity of 8 tonnes per hectare annually, a 1MW plant would require 800 hectares of land. India has a total land area of 328 million hectares, 45% of which is used for agriculture. A tree plantation to supply a 20,000 MW power plant would require 16 million hectares – about 5% of total land, or 12% of degraded land, in India.[30] If India were to generate all of its electricity needs from biomass, particularly firewood, then 1/4 to 1/3 of land would have to be used to grow wood. Together with agriculture, there would be hardly any area left for anything else, not to mention the complete destruction of biodiversity.

Sustainability results from more energy consumption

While the idea that increasing energy consumption increases sustainability may seem counterintuitive to some, energy consumption must increase to give humanity a lighter environmental footprint.

Energy consumption gives rise to energy efficiency, because those who consume more energy have an incentive to consume it more efficiently. This creates a huge market for efficiency measures, and triggers inventions and innovations that enable us all to become more energy efficient. The larger sum of energy consumed also gives incentives to substitute towards more efficient fuels. When energy is consumed in a larger chunk, it attracts users' attention, and persuades them to use it more efficiently.

When energy consumption is high, energy production and distribution can be achieved on a larger scale to reap the advantage of economies of scale and scope. Efficiency of energy production increases, and more kilowatt-hours of energy can be produced with the same tonnage of coal or oil. Increased efficiency in energy consumption also translates into reduced carbon intensity – economies emit less carbon to produce the same output. When energy consumption increases, people switch to more efficient fuels, which are also less carbon intensive. In contrast, initiatives that focus on sequestering carbon and cutting carbon emissions are likely to have the opposite effect. The world as a whole exhibits a pattern of 'decarbonisation' in its fuel use.

Countries such as the USA and Japan, where energy consumption levels are amongst the highest, are also the greatest innovators of energy-efficient technologies. When consumption levels are high, more efficient gadgets and devices come to the market and they quickly replace old and less energy-efficient devices. Alternatively, when consumption levels are low, these devices are irrelevant.

Barriers to large-scale energy provision in India

Generally, India lacks a sufficient infrastructure to facilitate widescale energy provision. Because of the extent of India's poverty there is also a lack of demand for energy – but of course, this is a vicious cycle. Generally, investors are apprehensive about investing in rural energy. Lack of financial and physical infrastructure, a lack of institutions (clearly defined property rights, enforceable contracts), corruption and regulations have worked to discourage commercial investors from investing in rural energy provision in India.

The Tata Energy Research Institute found that grid-based rural electrification programmes in India are largely unaffordable and unreliable, with an estimated cost of US$12,500 to US$30,000 per village, which means US$65–US$165 per household per year, depending on the distance from the existing grid.[31] Of course, if the opportunity costs of rural people are considered (see above), these figures may not be completely out of reach even for the very poor.

Nearly 88% of India's coal is produced and marketed by Coal India Limited and its subsidiaries. A consumer who requires coal must approach the Central Marketing Organisation. Due to bureaucratic hassles and the presence of a strong mafia in the coal sector, actual delivery is always at risk.

At the behest of international institutions, aid agencies and environmental groups, India's government has subsidised the renewable energy sector, seemingly forgetting about the need to generate affordable, reliable and plentiful energy for everyone.

Renewable providers enjoy guaranteed revenue streams and government protection from market signals. Meanwhile, subsidies for renewable energy use flow directly to high- or middle-income populations: 'Government largesse to renewable energy is comprehensive, widespread, and highly attractive ... A majority of wind energy projects in India have come up mainly to cash in on these tax breaks. Electric power generation is a secondary – and often neglected – priority.'[32]

Companies that invest in 'renewable' energy sources such as windmills get a 100%, one-year depreciation scheme from the Indian government, and a five-year income tax holiday. In Tamil Nadu during 1995–96, dozens of companies used this income tax break and accelerated depreciation to avoid paying any tax on income derived from other sources altogether.[33]

Thus cleaner and cheaper forms of energy, including gas, coal, hydro, oil and nuclear have been neglected. These forms of energy are far cheaper than solar and wind power in nearly all contexts. Moreover, they become cheaper as demand increases, which would happen as India's economy develops, and as companies take advantage of economies of scale.

Energy is a primary factor in industrial production. In India, State Electricity Boards (SEBs) control more than 85% of the total power generation, transmission and distribution. The SEBs are highly inefficient and because they are bankrupt, they cannot finance any addition to the installed capacity. Both peak load and energy shortages of varying degrees are prevalent in the country. This leads to perpetual scarcity, scheduled power cuts and outages.

However, India's federal and state governments have conspicuously failed to encourage an environment favourable towards private energy provision, by inhibiting development of new energy sources, over-regulating existing energy supplies, subsidising inefficient providers, providing monopoly power to certain providers, and generally by intervening in energy markets.

According to a 2002 study by the UK Department for International Development (DFID) called 'Energy for the Poor', the government of the Indian state of Andhra Pradesh was paying subsidies of US$600 million a year to the electricity board prior to power sector reform. The Indian minister for power indicated that in total SEBs lose the equivalent of US$9 billion a year.[34]

Wealthier household consumers insulate themselves from erratic electricity with diesel generators and inverters. Poorer consumers often simply do not have electricity. However, all over India poor entrepreneurs in the informal economy are getting around the problems created by the Indian state with their ingenuity and resourcefulness. Though this is very small-scale power generation, they are improving energy access for the poor – themselves – and have proven that poor people will pay for energy. This also illustrates that with formal markets,

---

Table 7 **India's commercial energy production and commercial energy per capita**[35]

Country	Commercial energy production (thousand metric tons of oil equivalent) 1999	Commercial energy use per capita (kg of oil equivalent) 1999
USA	1,687,886	8159
China	1,056,963	868
Bangladesh	14,474	139
India	409,788	482

---

consumers would likely benefit from lower prices due to economies of scale.

The poor and unreliable power supplied by SEBs also forces industries to install their own generating capacity because they cannot get enough energy to run their plants and factories.

Captive power is comparatively uneconomical and it is very cumbersome to acquire permission for setting up a captive plant. Most of the industries have to limit their production in absence of adequate power. Even India's per capita commercial energy use lags far behind other countries. It consumes about 55% of that consumed by China, and about 6% of that in the USA.

Another important factor in the lack of progress towards large-scale energy projects is opposition to hydrological power generation. Dam construction has both positive and negative consequences, but a small group of vocal opponents to dams has succeeded in stalling or abolishing proposals for new dams in India. Political tensions about water and power distribution, especially between states, have also contributed to India's lack of hydro-generating power.

Lastly, there are intrinsic regulations and cultural biases (which often become political) that ultimately affect access to energy. According to the DFID study:

> In many countries, the lack of legal status of poor people is a
> barrier to them having access to adequate energy services, even if

they can afford to pay for them. For example, migrants that move to shanty towns are often not allowed to be connected to the grid as they are not legally registered. Governments may be reluctant to recognise shanty towns as legal dwellings since they are then obliged to provide them with water and other infrastructure services ... In China, rural households that move to urban areas do not have 'urban status' and are therefore not allowed to be connected to power supplies.[36]

The UN's Clean Development Mechanism

In 1997, the Kyoto Protocol established the Clean Development Mechanism (CDM), which enables Annex I countries (developed countries and economies in transition) of the United Nations Framework Convention on Climate Change to flexibly meet their greenhouse gas reduction targets at a lower cost through projects in poor countries. The CDM is based on two complementary goals – to reduce the consumption of GHGs by reducing emissions, and to help poor countries with technology transfer:

> The purpose of the clean development mechanism shall be to assist Parties not included in Annex I in achieving sustainable development and in contributing to the ultimate objective of the Convention, and to assist Parties included in Annex I in achieving compliance with their quantified emission limitation and reduction commitments under Article 3.[37]

The CDM intends to allow wealthy countries to reach their agreed emissions reductions with a degree of flexibility, through voluntary partnering with poor countries.

> Article 12 of the Kyoto Protocol identifies three specific goals for the CDM: (1) to assist in the achievement of sustainable development, (2) to contribute to the attainment of the environmental goals of the Framework Convention, and (3) to assist Annex B parties in complying with their emissions reduction commitments.[38]

But the CDM's critics, including environmental organisations such as

WWF, suggest that it will do nothing to actually decrease carbon emissions:

> By intention, the CDM is not designed to reduce global
> greenhouse gas emissions. CDM projects that reduce emissions in
> the host countries will generate emissions credits that enable the
> investor countries to increase their domestic emissions, exceeding
> their Annex B emissions targets. Thus, at best, if the CDM
> operates as intended, it will be carbon-neutral on a global scale.
> However, in practice, to the extent that the CDM generates
> unwarranted free-rider credits, it will cause a net increase in global
> carbon emissions.[39]

As it is currently designed, the CDM is likely to be yet another UN-inspired bureaucracy which helps a few businesses and elite politicians at everyone else's expense. Projects under its auspices will be subjected to an endless number of criteria before approval, and it favours only large projects.

Because the CDM requires that potential projects fulfil so many criteria, it seems likely that the only projects it will attract are 1) those which would not be viable without government subsidies, or 2) those which have the best lobbying ability. However, it is unlikely that the UN will have the capacity to assess the opportunity cost of these projects against other economic activities which would take place in the absence of this 'crowding out' effect.

The CDM suffers from a broad problem which plagues many United Nations initiatives, which is that it is outcome-oriented rather than process-oriented. The broad outcome is to meet targets for reductions in greenhouse gas emissions – even though it is unclear that these targets are desirable, or will be helpful in combating global warming. Although it is intended to grant some amount of flexibility to the process, simply reaching an arbitrary target does not mean that we are progressing towards a broad goal of eliminating poverty and sustaining development.

This needless focus on targets and outcomes means that the CDM starts from the incorrect assumption that people are consuming too much energy. To that end, it proposes to restrict the availability and supply of energy, favours certain energy sources at the exclusion of others, and favours the development of certain new technologies.

Its preference for large scale projects means that the CDM neglects the most urgent human and environmental problem of poverty. For instance, if India were to reduce its energy consumption, this would only make poverty worse. One model estimates that over a period of 35 years, a 30% annual reduction in carbon emissions would lead to a decrease in GDP by 4%, and would increase the number of poor people by 17.5% in the thirtieth year.[40]

India urgently needs to reduce and eliminate poverty, not aggravate it by eliminating energy options. Eliminating poverty will only be possible by increasing the consumption of energy, especially for poor people. This would result in a virtuous cycle of development – cleaner, more efficient energies would lead to less degradation and more efficient use of resources. Ultimately, this would lead to conservation and efficiency in the energy sector. Wealthy countries have illustrated that this path of development works, and poor countries should not be discouraged from following it as well.

Technology transfer

Many commentators have praised the 'technology transfer' element of the CDM, suggesting that this is a superior way to assist development in poor countries.

Of course, new ideas and technologies are exchanged regardless of whether the UN intervenes in the transfer – this is an inherent part of the business cycle and of human adaptation. This dynamic process occurs especially when people can openly trade with each other, and when governments step back from dictating the process and outcomes of trade and investment.

Public and consumer demand also motivates technology transfer. In most developing countries today, consumers are widely adopting mobile phones instead of relying on arcane state-run telephone systems which barely work. The same will occur with clean energy technology, especially if governments concede that they cannot provide such services as well as the private sector.

Of course, the CDM may simply be yet another form of corporate welfare for businesses in wealthy countries to provide their technologies to poor countries. This could cause many negative effects, including undermining local incentives and initiatives to innovate and invest in new technologies that will satisfy the needs of local consumers. By subsidising certain technologies over others, the CDM may also 'crowd out' alternative sources of private sector investment in the energy sector, negating development and investment initiatives which would have otherwise taken place.

Of course, private sector businesses and entrepreneurs do not need subsidies to provide energy and technology to consumers. If governments and international agencies step out of the way, businesses – both local, national and international – will achieve this in a manner which satisfies consumers' needs and desires far better than governments, who have an extremely poor track record in this area.

Conclusion

Reliance on traditional energy has many disadvantages and is responsible for enormous human suffering, loss of life, reduced economic productivity, and environmental problems. Rural people have few other options, though, because governments in poor countries have conspicuously failed to facilitate broad access to energy.

For India's poor rural people, efficient, reliable energy remains a dream rather than a reality. Sadly, they will probably continue to suffer because of an unnecessary focus on reducing the amount of energy consumed in the world, to prevent the hypothetical, long-run risk of climate change. The real risk today is that billions of people in the world will not have the same opportunities to grow and develop as the First World did.

The immediate need of poor people in India and other poor countries is to consume more energy, in any form. Likewise, poor countries need plentiful, affordable, reliable and accessible energy of any kind – whether gas, nuclear, coal, oil, wind or solar, whether grid-based, locally generated or stand-alone – to fuel economic growth and improved quality of life for their people. More energy consumption will lead to more energy efficiency, which will lead to environmental benefits and sustainable energy consumption.

In developing countries, economic development usually means higher energy consumption, and more energy consumption temporarily means more air pollution. But in the long run, ever-wealthier developing countries will be able to improve their energy efficiency and thus also improve their air quality. In addition, pollution levels in the developing countries will not necessarily reach levels found earlier in rich countries during their development period. In the present era of globalisation, developing countries can gain access to modern technology quite easily. So the developing countries can achieve a high level of economic as well as environmental development in a much shorter period than did developed countries.

A truly 'clean' path of development would involve little government, and instead would rely on the initiative and ingenuity of people to solve energy needs. All over India, people in the informal economy are using their own ingenuity to achieve access to energy, and poor people are willing to pay for it.

The DFID study cited earlier correctly proposes that: 'Few people who have the interests of poor people at heart would advocate the maintenance of many of the current energy systems that are badly managed, deeply corrupt and suck in vast amounts of public money to underwrite huge and recurring losses.'[41]

Since India's government and most other poor countries have so clearly failed to improve access to energy for poor people, they should not be in the business of energy provision, or granting monopolies to certain providers, or enforcing burdensome regulations against those who would provide affordable energy.

Instead, governments should focus their energies towards establishing legal regimes which are transparent and uphold the rule of law, so that individuals and businesses can act on a level playing field to fulfil people's energy needs. This, in turn, would encourage adaptation to change – whether changes resulting from global warming, or any other phenomenon.

The Clean Development Mechanism, part of the Kyoto Protocol, has been promoted as a means to achieve reductions in greenhouse gas emissions while promoting technology transfer. However, because it is motivated by the wrong goal – one of limiting energy rather than broadly making energy available, affordable and clean for everyone – it is likely to hinder rather than promote sustainable development.

Addressing the lack of energy in poor rural areas should be a priority for those who care about eliminating poverty and promoting human well-being. Unless we solve the problems of poverty first, greenhouse gas mitigation strategies will not only be futile, but will heartlessly leave behind billions of people who today lead lives of drudgery and darkness.

6 Energy for the poor? The Clean Development Mechanism

Andrew Kenny

Introduction

Just over forty years ago in Africa, a man of vision announced his solution for the problems of poor black people. Noting that an ancient African culture was being degraded by an aggressive Western one, and that graceful traditions were being perverted by the trash and glitter of Western consumerism, this man proposed a grand scheme to preserve the old and simple ways of Africa. He wanted Africans to live close to nature in small communities, self-reliant in their modest needs. He wanted them to be protected from the corrupting influences of Western technology and Western avarice. He put his scheme into practice.

The man was former prime minister Dr Hendrik Verwoerd, the country was South Africa and the vision was called 'apartheid'. It resulted in repression, humiliation and mass poverty for millions of black people, and degradation of the environment.

A central theme of Verwoerd and the supporters of apartheid was that 'modern technology and wealth is fine for us whites but it will corrupt those poor blacks.' White supporters of apartheid told me, 'Your native doesn't want what you want. All he wants is a mud hut, three fat wives, a patch of mielies (maize) and a few cows.' Motor cars, air travel, electricity, flushing lavatories discharging into central sewers, brick houses with clean running water – we whites all have these things but those blacks should not have them.

Economist Robert H. Nelson of the University of Maryland has argued that 'The greatest current efforts to "save" Africa are associated with contemporary environmentalism. The results have not been as devastating as the experience of slavery, yet they have often served

Western interests and goals much more than the interests of ordinary Africans.'[1]

This chapter argues that the ideas of today's environmental groups have an astonishing and frightening similarity to those of Verwoerd and his followers, with negative implications for development in Africa. The spectre of global warming is used to encourage poor countries to sign up to agreements which will limit their energy consumption and perpetuate poverty – some of these sentiments are reflected in the UN's Clean Development Mechanism. But poor people do not want a new eco-imperialism: they want to grow wealthy, and energy is fundamental to wealth creation.

Resource sustainability

I attended a discussion on climate change at the University of Cape Town where somebody asked: 'Do you think everybody in the world could have the same living standards as the people in California now?' There was a frisson of horror from the green audience. The speaker replied with great embarrassment that it would not be possible with existing technology. Others have said that there are not enough resources. Both are quite wrong.

The same technology that has made Californians wealthy is available to anybody who wants to use it, and technology will improve, as it always does. The world has vastly more than enough resources to give everyone the same wealth as Californians and to sustain it indefinitely.[2] Almost every commodity needed in a modern economy is becoming cheaper and cheaper, more and more plentiful, and this trend will continue indefinitely.

A more crucial question, indeed the most crucial question in debate about man and the environment, is this: 'Do you think everybody in the world should have the same living standards as the people in California now?' It is necessary for the happiness of mankind and the health of our planet to answer 'yes'.

Poverty is bad for man and it is the greatest threat to the environment. Poor people breed more than rich, pollute rivers and local water supplies because they lack proper sanitation, cut down trees for firewood because they lack modern energy sources, plough down indigenous forest to practise primitive, low-yield agriculture, and overgraze and cause erosion with inefficient herding. Rich people lead lives which are more environmentally benign. They can afford cleaner and more efficient technologies, they produce more food on less land (and even produce too much food because of subsidies) and are secure enough in their everyday needs to worry about other priorities – caring for nature and regarding wildlife as a splendour rather than as a threat or a meal.

Lions once roamed through the whole of Europe. There are NO wild lions in Europe now. They were wiped out by desperately poor hunter-gatherers and primitive farmers in ancient times. Today their rich descendants pay a fortune to fly to Africa and take photographs of lions – and by so doing are helping to preserve them. But to save all of Africa's wildlife, our priority should be people, and specifically, to make African people wealthy as soon as possible.

Energy consumption is a fundamental requirement in all economies, and is essential to the development of poor countries. ('Poor countries' are sometimes condescendingly called the 'developing countries' or worse, 'the Third World', or worse still 'the Global South'.) Climate change now looms over all official considerations of energy use at both international and national levels, and it is casting a shadow over the energy policies of poor countries, who are being encouraged to buy into agreements which will ultimately limit their energy consumption.

Africans and other people with dark skins will simply be the victims of the decisions of rich white people on climate change, and the question 'Is climate change a real danger?' is not one they will be allowed to answer. But let me briefly address it.

Thirty years ago, the big scare was global cooling. We were urged by environmentalists to be terrified of a coming ice age. Nigel Calder, former editor of the *New Scientist*, wrote in *International Wildlife* in July 1975: 'The facts have emerged, in recent years and months, from research into past ice ages. They imply that the threat of a new ice age must now stand alongside nuclear war as a likely source of wholesale death and misery for mankind.'[3]

Other scientists and green commentators joined in, warning about plunging temperatures. As recently as January 1994, the supreme authority on matters environmental, *Time Magazine*, wrote:

The ice age cometh? Last week's big chill was a reminder that the Earth's climate can change at any time ... The last one [ice age]

ended 10,000 years ago; the next one – for there will be a next
one – could start tens of thousands of years from now. Or tens of
years. Or it may have already started ... Temperatures in dozens
of US cities dropped to all-time lows ... Chicago schools closed
because of cold weather for the first time in history ... the city's
lows were below -23°C for a record 10 straight days.[4]

And the fact is that alarmism sells magazines: we were all going to
freeze. Now we are all going to fry. All you need to do to convert from
one scare to another is to replace 'unprecedented cooling' with 'unprecedented warming'.

The facts are these. Carbon dioxide is a greenhouse gas (which
traps heat); carbon dioxide has been increasing in the atmosphere and
the levels are higher now than they have been in at least 400,000
years; and the increase is because of man.[5]

The rest is guesswork. We do not understand the Earth's climate
system. We can draw no conclusions from temperature records (temperatures in northern Europe were much higher a thousand years ago
than they are now, and this led to a boom in agriculture).[6] Above all,
we do not know what causes ice ages, which have occurred in
100,000-year cycles over the last 2 million years but not before. A
new ice age, for which we are due, would be an unmitigated disaster.
Ice ages happen when carbon dioxide concentrations in the atmosphere are low, but whether this is a cause or an effect is unknown.

In the face of this lack of understanding, how should we apply the
precautionary principle, which suggests that we act to eliminate all
potential threats? As the greatest danger seems to be an imminent ice
age, should we release as much carbon dioxide as we can, giving
power stations and industry tax breaks for every ton they release?
Should we pretend that ice ages do not happen, and reduce carbon
dioxide as much as we can, based on some speculation about global
warming? Or should we be honest about our ignorance and do nothing at all?

'Clean' development

In practice the science does not matter since those with the power and
the influence have decreed that global warming is occurring, its effects
will be bad, it is caused by man's reliance on hydrocarbon fuels, and

it must be resisted and halted by cutting greenhouse gases. Europeans bemusedly watched the workings of African superstitions. Africans bemusedly watched while Europeans feuded first over their Christian factions and then over their Cold War ideologies. But African beliefs hardly touched Europe, whereas European beliefs rocked Africa. And now the white missionary has a new religion – climate change – and is offering his dark-skinned flock a new rite – the Clean Development Mechanism (CDM).

The 1997 Kyoto Protocol has legally binding targets for rich countries (termed Annex 1 countries) to reduce their greenhouse emissions in total by at least 5% below 1990 levels during the period 2008 to 2012. The poor countries (termed Non-Annex 1 countries) have no obligations – thank goodness. But to encourage them to reduce greenhouse gas emissions, a mechanism is being proposed to allow for the rich to pay the poor to reduce emissions.

The CDM will allow rich countries to meet their own emissions cuts by reducing emissions in a poor country. If a rich country pays a poor country to reduce its carbon dioxide emissions by one ton, the rich country can claim that ton as credit towards meeting its own Kyoto targets.

One of the premises of CDM is that it is easier to reduce greenhouse emissions with the primitive technologies of the poor countries than with the advanced technologies of wealthy countries. This is perfectly valid: a dollar spent improving the efficiency of an efficient machine will yield less return than a dollar spent improving the efficiency of an inefficient one. It is much easier to reduce the carbon dioxide released when an African woman in a township cooks her evening meal over a coal fire than when a Parisian woman cooks hers over an electric stove powered by a nuclear station.

The CDM pledges that if poor countries adopt 'clean technologies' ('clean' according to this definition means technologies which reduce greenhouse gas emissions) they will be paid by the rich countries. This will happen in a complicated scheme of 'carbon trading' and 'Certified Emission Reductions' backed by a huge bureaucracy which will 'authorise, validate and register' projects, establish 'baselines' and do a lot of 'monitoring' to measure 'avoided emissions'. There is talk of a 'carbon economy', some of whose present instruments are 'The World Bank Prototype Carbon Fund', 'The Dutch Carbon Credits Purchase' and the 'UK Emissions Trading Scheme'.[7]

The idea of selling the pollution is probably a good one – it will probably achieve environmental protection more efficiently and with better results than penalties or regulated limits for pollution. Carbon dioxide is not a pollutant in the strict sense, but the same arguments hold for it. So, a market in permits to emit carbon dioxide, which could operate within and between countries, has merits.

However, the complexity of the CDM mechanism looks daunting and the prospect of a vast international army of inspectors and monitors is not appealing to most. The most ominous feature of the system, though, is the possibility of choosing entirely unsuitable CDM projects based more on the interests and ideology of the wealthy supplier than of the poor recipient.

Africa is littered with relics of the white man's folly. The Tanganyika Groundnuts Scheme, Nyerere's Ujaama socialist farms heavily financed by the West, solar power installations that never worked, and myriad other projects and schemes that crumbled the moment they were implemented all bear testimony to an arrogant stupidity from the white sponsors. Of course, corrupt black leaders built lavish conference halls, grand palaces and huge international airports while their economies collapsed. They are also to blame, but most of their extravaganzas would not have been possible without Western aid or loan money.

There are two keys to the success of any project in Africa. The first is to understand what African people want rather than what African leaders or white donors want. The second is to have objective measures of cost and benefit.

It is no surprise, contrary to the ideas of Verwoerd and his supporters, that black people nearly always want the same things as white people. To begin with, most would prefer to live in cities and suburbs than in the countryside.

Urbanisation is a universal trend around the world. According to people who have experienced both, a slum in the city is better than a village in the hills. So, in Asia, Latin America and Africa, they pour into town. This horrifies green ideologues in exactly the same way that it horrified Dr Verwoerd. Indeed, the single greatest battle of apartheid, fought with the utmost brutality and complete failure, was to stop rural black people coming into the 'white' cities.

But urbanisation is almost wholly good. It is much easier to improve access to services which improve human well-being – such as running water, sewerage, electricity, waste collection, transport, communications and education – to people in urban areas than in rural areas. It is urban areas which present more economic opportunities than the countryside.

And urbanisation, despite myths promoted by environmental groups, is beneficial for the environment. The greatest threat to African wildlife is the encroachment of poor farmers and the predations of poachers in the countryside. If people move to cities, it will be easier to preserve wild places, and the animals of Africa, a wonder of the world, will be left free and secure, visited only by game rangers, local enthusiasts and paying tourists from abroad. The best possible solution for Africa is a great area of wilderness filled with our planet's most magnificent fauna, an area of commercial farms feeding the continent efficiently and most of the human population living in cities and suburbs, which is just what they want to do.

As a start, there should be no CDM projects that try to force people to stay in the countryside. The CDM's projects should encourage urbanisation and help to make urban life safer, cleaner, healthier and more prosperous.

It is extremely important to have objective measurements of the costs and benefits of different energy options. Right now, such judgements seem to be based on emotion. For example, there is excessive anxiety about industrial pollution and very little on household pollution. But in South Africa, and no doubt in the rest of Africa, Latin America and Asia, the health hazards of fuels used for home cooking and lighting are much larger than those posed by big power stations or industry.

If you drive past the townships of South Africa on a still winter day, you will see an evil smog lying over them like the sheet over a dead man. This comes from the burning of wood, coal and paraffin in households, which kills and debilitates on a huge scale. In South Africa, the mortality rate for acute respiratory infections in children is 270 times greater than in western Europe; this is because of indoor air pollution caused by burning wood, coal and paraffin.[8] Electricity from the dirtiest possible power station (coal) provides energy which is hundreds of times cleaner and healthier than that from burning fuel inside a shanty.

The single greatest energy need for an African woman is the energy to cook an evening meal in the middle of winter. In the countryside, she cooks it with wood or dung for fuel. She might spend three hours collecting the wood and she might chop down trees to get it, adding to Africa's land degradation, which has happened in Dr Verwoerd's model Bantustan, the Transkei. Wood, which is 'renewable', is more likely to suffer depletion through abuse than non-renewable energy such as coal, and can be extremely unhealthy as an energy source. In the townships, a woman cooks with coal, paraffin or LPG. If she is lucky, she will have electricity for cooking. No other energy decision has more effect on life and death than this African woman selecting the fuel for her evening meal.

A colleague prepared a brilliant slide on the comparative perils of energy.[9] The slide is called 'two paraffin accidents'. It is divided into two sides. The right side, with the lesser accident, shows New York's World Trade Center buildings on fire on 11 September 2001. The explosions and fire which caused this tragedy were ignited by jet fuel, which is paraffin. It killed just under 3,000 people.

The left side, with the major accident, shows a cheap paraffin stove knocked over and in flames. This sort of cooking stove is used by African people in the townships. In South Africa alone, it causes over twice as many deaths every year as the number caused by September 11. The stoves are badly designed, so that the paraffin in the reservoir is heated to ignition temperature by the burner and, if the stoves are bumped or knocked over, paraffin spills out and explodes.

The result is death and disability on a massive scale. Week after week, fires caused by these explosions rip through the tinder constructions of the shacks in the squatter campers and townships. Thousands of people die every year. Infants who survive the flames are often left with their faces burned off and spend the rest of their lives as monsters. To add to the horror, paraffin is highly toxic but colourless, and usually stored in beverage bottles. Infants often drink it – and die. Paraffin poisoning kills over 4,000 children a year in South Africa.[10] Finally, the emissions from paraffin burning in these stoves are dangerous to health.

A far safer fuel is liquid petroleum gas (LPG). It is stored in gas bottles and so there is no chance of accidental ingestion. It burns cleanly. LPG stoves do not explode if they are knocked over. LPG is two orders of magnitude safer than cheap paraffin stoves.[11] LPG is a product of oil refineries, plentifully available. Unfortunately, LPG stoves, like the ones used for camping, are considerably more expensive than paraffin stoves.

A wonderful CDM project would be to replace cheap paraffin stoves with LPG stoves. Because its ratio of hydrogen to carbon atoms is higher than that of paraffin and because LPG stoves cook more efficiently than cheap paraffin ones, LPG releases less carbon dioxide for every joule of useful energy, and so would probably qualify for CDM. A project financed by a Western donor to design and make a cheaper LPG stove, and to set up an efficient distribution and marketing system for LPG in the rural areas and townships, would do more to save African lives and improve African health than any other energy project.

What would not benefit Africa are daft 'renewable energy' projects. In almost all cases, wind and solar energy is useless for Africa. It is very expensive (relative to other energy sources), unreliable, fragile, difficult to maintain and does not provide energy when you want it – for example, to cook the evening meal in the middle of winter. This means that renewable energy for Africa is usually unsustainable – though there are a few exceptions, such as water heating. For making electricity, solar and wind are extremely costly and so can only be used in poor communities for generating tiny amounts, suitable for lighting and radios (both important) but not for heating or cooking.

There is a simple-minded attraction to the idea that Africa is hot and sunny and therefore solar power is a good thing for it. But even in Africa, sunshine is very dilute and intermittent, and can only be harnessed in useful quantities at great cost and low reliability.

I gained insight into the social consequences of solar electricity when an American worker in 'alternative energy' came to Africa. She told us about a survey she had conducted among African villagers, asking them whether they wanted electricity and what they wanted it for. Everyone wanted it but there was a stark gender difference in what they wanted it for. The women wanted it for cooking and heating. The men wanted it for entertainment (radio, television and CD players). Since solar power can only deliver tiny amounts of electricity at reasonable costs, it was useless for the first but acceptable for the second. She concluded, 'Solar power is a very guy thing.'

As an engineering student, I had to design a solar-powered refrigerator for storing vaccines in remote African clinics. This seemed a rather good idea because refrigeration is most needed when the sun is providing the most energy. I approached a refrigeration contractor who had had decades of experience in solar power in Africa, and asked him the best way to run a small refrigerator in the heart of the bush. Without hesitation he replied: 'A diesel generator.'

Unfortunately, though, there are European companies who are looking to make a killing by peddling wind turbines and solar power equipment to the poor countries under CDM. More important, wealthy environmental ideologues are besotted with wind and solar. Their experience with such energy is usually limited to recreational experiences – the photovoltaic panel on their yachts, the charming little wind generator in their safari camp in Kenya, and they see no reason why black chappies should not have them all the time. In rich countries with energy intensive economies, such as Denmark, renewable electricity is heavily subsidised as a sort of self-indulgence. If these patronising and entirely wrong sentiments become the driving force for CDM projects on solar and wind electricity generation, Africa will suffer.

South Africa has embarked on the world's most ambitious programme to electrify poor communities.[12] Grid electricity is brought first to those living closest to the existing grid. Those far from it are offered photovoltaic panels, and if they accept, they are less likely to be connected to the grid.

The experience of the two groups has rapidly been communicated throughout the country. When black people in villages are approached and asked if they would like solar panels for electricity, they always reject them in dismay. They do not want the 'weak' electricity from solar panels, they want the 'strong' electricity from the grid. In some provinces, there are subsidised schemes for private power companies to supply villagers with a combination of photovoltaic units for lighting and radio, and LPG for cooking and heating. This is a sensible compromise, and would also be a good candidate for CDM. But it is only an interim measure until the villagers get grid connection either to the central grid or a local one.

All of humanity is moving towards reliance on electricity as the best form of energy. Electricity is clean, convenient, versatile, superbly ordered and quite safe. Lenin was right about one thing at least: electrification benefits man. With the exception of energy for heat, which may be better provided by other sources, electricity is the optimum energy. All attempts to bring electricity to the poor are motivated by good intentions even though they may fail in practice (mainly because poor people cannot afford to pay their electricity bills). Regardless, to make the poor countries rich, they must encourage a variety of solutions to energy provision, including large, grid-based electricity.

So which is the best source of electrical energy, which will most reduce greenhouse gases, with the best safety record, and is most environmentally benign? It is on this question that CDM fails worst. By every objective measure of safety, health, economics and the environment, the best source of energy for making electricity is nuclear power.

Nuclear power has an unrivalled safety record. No other source of large-scale energy comes close. Nuclear power currently provides 17% of the world's electricity. The worst ever nuclear power station accident in the West, during over 40 years of experience, was at Three Mile Island in 1979. It killed no one, injured no one and had no ill health effects afterwards. According to the Paul Scherrer Institut, the number of accidents in the energy sector between 1969 and 1996 which killed at least five people was the following: coal – 187, oil – 334, natural gas – 86, LPG – 77, hydropower – 9, nuclear – 1.[13] The single nuclear accident was Chernobyl, whose primary cause was a mad reactor design that would never have been allowed in the West.

There is very little connection between nuclear power and nuclear weapons. Weapons require enrichment over 90%; nuclear power uses enrichment under 10%. From nuclear power reactors that run for more than a month without waste removal – the great majority – their waste is useless for making weapons.[14]

In operation, nuclear power reactors release no greenhouse gases, nor any other air pollution. Over the whole energy cycle, including construction, fuel preparation, operation and decommissioning, nuclear power releases amongst the fewest greenhouse gases of any energy source, including wind and solar power.[15] The radiation from nuclear stations is tiny, less than that from coal power stations and much less than that from large hospitals.

Above all, nuclear power has the least waste problem, producing a tiny amount of radioactive waste which is solid, stable and easy to store so that it presents no danger to man or the environment. By contrast, the waste from coal stations is massively larger, far more dangerous and lasts much longer. Coal waste includes heavy metal toxins such as mercury and arsenic, which remain dangerous forever, and radioactive elements such as thorium, which has a half-life of 14 billion years. This is simply hurled into the air we breathe or dumped on to ash tips – with never a peep of protest from anti-nuclear groups who fret about the problem of nuclear waste.

But nuclear power is specifically ruled out for CDM, which gives reason to question its intentions and goals. Does the CDM intend to reduce greenhouse gases through clean and safe technologies? Or does it intend to promote an irrational green ideology, which regards nuclear power as sixteenth-century witch-finders regarded witches? Many of the non-Annex 1 countries already have nuclear power stations which run successfully, safely and efficiently. These include India, China, South Korea, Taiwan and South Africa. It is illogical and outrageous that their new nuclear power stations do not qualify for CDM.

South Africa, which has a highly energy-intensive economy, releases about 300 million tons of carbon dioxide equivalent a year, half of which comes from coal-fired power stations.[16] South Africa produces over 90% of its electricity from coal. The coal stations do not have flue gas desulphurisation because the public utility, Eskom, decided that cheaper electricity was a bigger benefit than the slightly cleaner air you would get by paying for expensive desulphurisation equipment. It was right.

By building nuclear stations in the future, South Africa would reduce greenhouse gas emissions more than the whole continent's wind and solar projects put together – probably more than the world's wind and solar projects put together. And it so happens that South Africa is developing a new nuclear reactor, the Pebble Bed Modular Reactor (PBMR), based on a proven German design. Its design philosophy is inherent or passive safety. No matter what human error or equipment failure, it is impossible to have an accident that endangers the public. It is small, simple and cheap. In an honest world, this would be a prime candidate for CDM.

Africa has other potential low-emission energy sources. For instance, it has the world's biggest untapped potential for hydropower. One site on the Congo River alone, at the Inga Falls, could provide up to 100,000 MWe – twice the electricity consumption of the British Isles or twice that of the African continent. And this would generate electricity through the flow of the river, rather than with a dam. But there is an ideological objection to all hydropower from the environmental groups. Hydropower at Inga Falls would not need a dam at all, so it would have almost no environmental consequences. Hydropower should also be a prime target for CDM projects.

Many African countries have failing power stations and electricity grids. Zimbabwe is a good example. It has more than enough generation capacity to meet its needs, but it keeps running out of electricity. The reasons are purely political: competent electricity managers were replaced with incompetent political cronies, and a loss of foreign exchange thanks to President Mugabe's ruinous policies has made it difficult to get equipment to maintain the power stations, coal mines and distribution networks.

If calculations showed that the extra electricity obtained from repairing existing African grids would release less carbon dioxide than the energy it would replace (for example if electricity replaced candles and coal as household fuel), then an excellent CDM project would be simply to refurbish the power stations and distribution systems and get them working again. Of course, political reform is essential but CDM, which is also highly political and bristling with administrative procedures and directives, could actually be useful in working side by side with the African bureaucracies.

In 1900, average human life expectancy in the world was 30 years. In 2000, it was 60.[17] The main reason for the improvement was the big five benefits of the Industrial Revolution: a brick house, clean running water, good sanitation, decent food and electricity. The lesser reason was medicine, especially in combating infectious diseases. The Industrial Revolution was based on the energy from fossil fuels. Development for the poor countries using fossil fuels, as Europe did, is incomparably better than no development at all – better for man and the environment. To achieve the same development using fewer hydrocarbon fuels is possible, because we have substantially better technologies now than in the nineteenth century. But our energy sources should be selected carefully and objectively, based on calculations of cost and benefit.

If representatives of wealthy countries – namely, green NGOs, donor agencies, and international agencies – insist on 'bringing the truth to the natives', this time in the form of the Clean Development Mechanism, they should know what the truth is. Poor countries do not want a new eco-imperialism, or a sequel to Verwoerd's apartheid, in the form of windmills that hardly produce enough electricity to make a piece of toast, solar cookers that can only cook in the middle of the day if there are no clouds, or tanks of fermenting pig waste that need a porcine multitude to supply them. These are hopelessly inappropriate, deeply patronising schemes of the wealthy green elite who think that energy comes from electric outlets or gas taps.

If wealthy countries really want to help the poor, the best thing they can do is to trade with them freely, getting rid of the wicked subsidies to wealthy farmers and eliminating protectionism in agriculture, textiles and other industries where the poor countries are highly competitive.

Poor countries are poor because they have not experienced their own Industrial Revolution, which drove wealth creation. People in poor countries want to achieve prosperity, literacy, efficiency, health and well-being – and rich countries achieved this through wealth.

Part Three

The Economic Consequences of Climate Change Policy

7 Warming aid, chilling trade?

Julian Morris

Global warming is the mother of all environmental scares. In the scope of its consequences for life on planet earth and the immense size of its remedies, global warming dwarfs all the other environmental and safety scares of our time put together. Warming (and warming alone), through its primary antidote of withdrawing carbon from production and consumption, is capable of realizing the environmentalists' dream of an egalitarian society based on rejection of economic growth in favour of a smaller population eating lower on the food chain, consuming a lot less, and sharing a much lower level of resources much more equally ...

[The environmentalists'] favoured political mechanism [for achieving this outcome] would be an international treaty ... Think (if you are persuaded you hold the truth) of the glory of it: no need to cope with regulations in different countries ... Everything can be done uniformly and worldwide by central direction.

Aaron Wildavsky, 1992[1]

When Aaron Wildavsky wrote this passage, I was myself a naïve young environmentalist. I had just written a paper on how to establish a global system for restricting emissions of carbon dioxide and was about to embark on a Masters course in Environment and Resource Economics run by Professor David Pearce at University College London. At the time, there was much buzz about a conference due to take place in Rio de Janeiro that was being hailed as the 'Earth Summit' and global warming was a hot item on the agenda.

Eleven years later, much has changed. The international establishment, including environmentalists, has achieved many of its goals in advancing regulations to deal with the threat of global warming. They have established a set of institutions, including an international treaty, the Kyoto Protocol, through which they would control world energy resources. They have bullied two of the world's three largest oil producers into accepting the basic proposition that global control of energy resources is desirable. They have scared many people into believing that absurdly high taxes on fuel are good for the environment. And they have likewise persuaded many opinion formers around the globe of the need for more public transport, renewable energy and other supposedly carbon-reducing fixes for global warming.

However, several obstacles have prevented the environmentalists' dream from becoming a reality. Specifically, the USA, Australia and Russia have not ratified the Kyoto Protocol.[2] Without ratification by at least one of these governments, the Protocol will not enter into force. Even if one of them (most likely Russia) does ratify, two of the world's largest producers of carbon dioxide are not members of the treaty and hence cannot have their energy use dictated by the environmentalist kings. In addition, numerous governments in poorer parts of the world seem ambivalent at best about implementing the Kyoto Protocol.

In order to overcome these obstacles the environmentalists have devised a cunning plan.[3] This involves a combination of aid to poor country governments and the use of trade sanctions against non-compliant rich country governments. In addition, a coalition of businesses has been pushing for implementation of the Kyoto Protocol, combined with trade sanctions as a means of protecting themselves from international competition.

The purpose of this chapter is to discuss the mechanisms available for the enforcement of the Kyoto Protocol and specifically the potential role of trade sanctions as a means of enforcement.

Enforcement of the Kyoto Protocol

Multilateral Environmental Agreements (MEAs) such as the Kyoto Protocol are notoriously difficult to enforce. There are two main reasons for this: the first relates to distribution of benefits and costs; the second relates to the willingness of the parties to the convention to utilise effective mechanisms of compliance. I shall address each in turn.

Some MEAs generate relatively large benefits and relatively small costs for nearly every party. Some have argued that this is true for the Vienna Convention on Ozone Depleting Substances, which established a framework for reducing the production of CFCs that are blamed for causing stratospheric ozone depletion.[4] Under such circumstances, it has been observed, the agreement is essentially self-enforcing: most parties have an incentive to comply and do so without any external encouragement.[5] The main role of the agreement in this case is to specify an equitable distribution of costs.

Other MEAs create significant benefits and relatively few costs for their members, but the costs to non-members are far greater than the benefits. The North Pacific Fur Seal Treaty of 1911 sought to conserve seal stocks that traversed the high seas. The treaty had four members: the UK (on behalf of Canada), the USA, Japan and Russia. This small membership and the significant individual benefit of compliance made this treaty self-enforcing among members. However, it was not self-enforcing against non-members, who – in the absence of any enforcement mechanism – might have hunted seals, undermining the effectiveness of the treaty.

To deal with this potential threat, parties threatened non-parties with trade sanctions. What the treaty did, in effect, was create property rights in these fur seals, transferring them from open access to the exclusive domain of the signatory parties. The trade sanctions then operated as a mechanism to enforce exclusion on others, just as legal penalties operate as an exclusion mechanism for the owners of other forms of property, discouraging trespass, nuisance and theft.

For a convention such as Kyoto, whose costs fall on one group of parties whilst the benefits fall on another group, a self-enforcing agreement is almost out of the question, so various methods of inducing ratification and enforcing compliance have been devised.

The use of aid to induce ratification

One way to overcome the problem of unequal distribution of benefits and costs resulting from the Kyoto Protocol is to offer financial transfers to those who would experience disproportionate net costs from the effects of global warming.

In the conventional analysis the main losers from climate change are assumed to be people in poor countries. So, logically, poor countries should be made to pay rich countries in order to compensate them for the losses the rich incur by reducing emissions. However, such a dispassionate economic analysis ignores the realities of Kyotonomics, which is predicated on egalitarian assumptions. Thus, since climate change is deemed to be the fault of rich countries, it is the rich countries that must pay compensation to poor countries on the premise that they might suffer in the future.

Of course, there is both a pragmatic and an ethical logic to this. The pragmatic element is that poor countries have other priorities: nearly a billion people in poor countries continue to live at subsistence levels, malnourished and suffering from frequent debilitating diseases; approximately 1.7 million people, most of them children, die every year in poor countries from diseases that result from poor water quality and a lack of sanitation; millions more die from respiratory infections and other preventable or curable diseases.[6]

Under such circumstances, signing up to a Protocol that will have little or no benefit for at least 50 years is even a waste of bureaucratic energy. This is significant, since bureaucrats usually have endless amounts of energy to spend in pursuit of mindless outcomes.

The ethical element is that since rich countries currently emit most of the carbon dioxide, it seems reasonable that they should have the responsibility of dealing with problems created by those emissions. But this presents an ethical dilemma: does the sin of emitting carbon dioxide necessitate the evil of paying bribes to Third World politicians in order to induce them to ratify an agreement? The answer of the dream-seeking environmentalist seems to be yes.

In 1992, when the UN Framework Convention on Climate Change (FCCC) was negotiated at the Earth Summit, poor country governments sought a *quid pro quo* for signing up and they got one. In return for their acceptance of the Rio Declaration, the FCCC, a declaration on forests and the Convention on Biological Diversity, poor countries were promised more aid.[7] African governments in particular were offered the carrot of a Convention to Combat Desertification (CCD), which contained various provisions for government-to-government financial transfers.[8] The FCCC itself reminds us that our obligations towards the climate exist 'on the basis of equity and in accordance with their common but differentiated responsibilities and respective capabilities'. Thus spake the environmentalist.

The promised aid was slow to materialise and much of it never did. Although some governments (or, more pertinently, some government officials in poor countries) benefited from monies from the Global Environment Facility, the sums involved were relatively small: approximately US$2 billion per year.[9]

Whilst these dribbles from the GEF and other agencies kept poor countries happy as regards ratifying the FCCC, when it came to implementing the Convention through an agreement on targets and timetables, rich country governments were forced to increase the size of the carrot on offer. That carrot came in the form of a mechanism for transferring technologies from rich to poor countries. The Clean Development Mechanism (see Chapters 5, 'Sustainable energy for the poor', and 6, 'Energy for the poor? The Clean Development Mechanism'), will allow rich countries to generate emissions credits for reducing emissions in poor countries. Sceptics, including many environmentalists, have noted that the CDM will do little to reduce carbon emissions. It will also do little to benefit poor countries; indeed it may harm them by locking them into a low-energy, low-growth development path.

The use of trade sanctions as an enforcement mechanism

Under some circumstances trade sanctions offer a means to enforce agreements. However, there have been relatively few instances where such sanctions have proved successful as enforcement mechanisms. Nevertheless, there are numerous MEAs containing explicit trade provisions, but most have been counterproductive.[10]

Consider the Convention on Trade in Endangered Species (CITES). Signed in 1973, CITES is predicated on the assumption that trade leads to the extinction of certain species. However, at least for the species that have been analysed in detail, this presumption is inaccurate. One reason is that CITES trade bans don't work. Since 1973, when a CITES ban was instituted, trade in rhino horn has not ceased and nor has the decline in rhino numbers slowed, except where those rhinos are well protected.[11]

This latter point demonstrates a more fundamental error in CITES, that it aims at the wrong target. The problem is not trade (if it was, then sheep and cows would also be extinct); it is the incentives to conserve. In the case of rhinos and most other land-dwelling species (such as the buffalo), where there are no property rights, individuals will have no incentive to invest in conservation and will instead simply consume what they want when they want it – until the resource is depleted. Trade may hasten the process by increasing the relative value of dead animals to live animals, but it is not the *cause*.

Experience shows that if animals are subject to well-defined and enforceable private property rights, then the people who own those rights will have an incentive to conserve the animals in order to maximise their economic value. Trade in this case can increase the return on investments in conservation by increasing the price paid for the products, thereby giving people an even stronger incentive to conserve.

Trade, the environment and the WTO[12]

Statutory limitations on imports and exports have been common throughout history, and generally have been used by the state to raise revenue and to bestow preferential treatment on favoured firms. They were often justified as a means of enhancing the wealth of a nation, based on the idea that exports are good and imports are bad. These arguments were demolished by Adam Smith (1776), who showed that trade produces mutual gains, and David Ricardo (1817), who showed that this applies to trade between nations even when there is no absolute cost advantage.[13] These insights still hold and have very clear implications for the environment.

First, the costs of producing goods vary due to the environmental conditions that pertain. By enabling trade to take place, production will tend to occur in places where the environmental conditions are more suitable.

Production of aluminium, which requires large amounts of electricity, occurs in areas where cheap electricity is available from hydroelectric power stations. This often entails transporting raw material – bauxite – thousands of miles, from one continent to another. Transporting bauxite clearly necessitates the consumption of more raw materials than transporting the finished product. However the important factor here is that producing aluminium closer to the bauxite mines would probably be less environmentally efficient, necessitating the use of more resources and the emission of more pollution to generate electricity than results from transporting the bauxite and producing aluminium using hydroelectric power.

In addition, open trade enables competition between suppliers, which drives technological innovation as suppliers seek to satisfy consumer demands by producing better products at a lower cost. New technologies themselves typically consume fewer resources for each unit of service they provide to consumers: fibreoptic cables carry thousands of times more information than their copper predecessors, and are produced with fewer resources. A small computer today can process vastly more information than computers the size of a house could process fifty years ago. At a more prosaic level, soft drink cans are now far lighter and as a result require fewer resources in production and distribution than they did twenty and thirty years ago.

New technologies also increase economic efficiency, raising productivity, which leads to economic growth. As a result, people become wealthier. And wealthier people are more willing to spend money on goods such as environmental protection, which tends to improve indefinitely as countries become wealthier. Wealthy countries have largely eliminated egregious environmental problems, and with economic development poor countries will too.

Thus, we may conclude that agreements limiting the use of tariff and non-tariff barriers to trade result in direct improvements to human welfare and the environment. We are enormously fortunate to live in an era when these barriers continue to fall. During the past fifty years a series of international agreements have reduced such barriers, first under the framework of the General Agreement on Tariffs and Trade (GATT) and then under the WTO (of which the GATT is now a part).

The GATT knocked down barriers to trade by establishing various principles. Perhaps the most important of these is the principle of non-discrimination, which requires *inter alia* that any restrictions must be applied equally to national producers and importers, and that they cannot be a disguised restraint on trade. In particular, there is a general prohibition on the use of trade sanctions to discriminate against the methods by which goods are produced when these do not affect the qualities of the good itself (this is called non-product-related process and production methods, or PPMs).

Evaluating policies towards trade and the environment

Given the potential benefits of trade liberalisation for human welfare and for the environment, Professor David Pearce has suggested that policies which restrict international trade for environmental reasons must pass three tests:

1   They must show that the environmental degradation brought about by free trade is a) truly brought about by trade rather than some other factor, and b) of greater consequence than the losses of human well-being that would ensure from restricted trade.

2   They must show that production-related damage is a legitimate feature of the importing nations' loss of well-being.

3   They must show that a trade restriction is the most cost-effective way of bringing about the change in the product or process which gives rise to the externality.[14]

On the first point, it has been observed above that trade is generally beneficial to the environment, so the burden of proof is well and truly on those who claim the opposite to be the case. When one assesses the claims made in specific cases by environmentalists regarding the impact of trade on the environment, the claims are often if not always found to be incorrect. For example, it has been widely claimed that species losses, from elephants to tropical hardwood, are being driven by trade. However the reality is that trade is not a major cause of such losses. The real culprit is inefficient and ineffectual systems of ownership and enforcement of ownership.

Similarly, the Basel Convention, which restricts trade in so-called hazardous waste between OECD and non-OECD countries, is predicated on the assumption that trade in such waste is harmful. But is this assumption valid? There are no doubt many instances of illegal dumping of waste in poor countries and these no doubt cause harm to people and the environment. However, the evidence suggests that the cause is lack of enforcement of laws locally, not the international trade in waste, and this is far less related to trade between rich and poor countries. Indeed, poor countries produce a considerable amount of waste locally, much of it potentially hazardous. It seems unlikely that the Basel Convention would improve local management of such waste. Indeed, it may make things worse. A voluntary restriction on the import of waste lead into India undermined the formal lead recycling industry in that country, which was dependent on high volumes of material. The result was an increase in informal recycling, which is more hazardous both to people and to the environment.

The environmental priorities of people in developing countries are significantly different from those of people in the developed world. For environmentalists in the developed world, global environmental standards mean less landfilling of waste and increased government protection of endangered species. It's not that people in poor countries care less about the environment. To them, though, such 'standards' seem absurd in light of their immediate needs and priorities – access to clean water, a reasonably reliable supply of electricity, and the ability to generate income (amongst other things), which would enable them to access new technologies.

From these two examples, it is clear that measures to restrict trade are rarely, if ever, necessary to achieve environmental goals. This is because environmental degradation is rarely, if ever, brought about by trade. It is usually the consequence of poor enforcement of laws at the local level and can best be addressed by improving those laws, not by imposing restrictions on trade.

Second, it is clear that such trade restrictions often have a human cost that outweighs any putative environmental cost. In the cases cited above, the loss of revenue to local people from the restriction on trade in wildlife, and the loss of jobs of people working in waste-related activities in poor countries would be examples. So, on this analysis trade restrictions in MEAs fail on both parts of the first hurdle set by Professor Pearce.

As for the second part, it is perhaps difficult to evaluate the extent to which the well-being of the importing country is improved by restricting goods on production-related grounds. This has been tested several times by the Dispute Settlement body of the World Trade Organization, with varied consequences. I now briefly describe the most important cases.

Tuna–Dolphin I
The first case involved a dispute over the import into the USA of tuna caught in ways that are allegedly harmful to dolphins. The USA had imposed restrictions on tuna caught using purse seine nets (which can trap dolphins), on the basis that its own tuna fishermen were governed by the same restrictions under a US law, the Marine Mammal Protection Act (MMPA).

The Dispute Panel of GATT, which at the time was the main system which governed world trade, reported that application of the MMPA to Mexican tuna constituted a breach of Article XI(1) of the GATT, which says that tariffs are the only legitimate means to limit imports and that other restrictions are unjustified restraints on trade. The ban on tuna imports was said to be a 'quantitative restriction' and therefore illegal.

The panel noted that internal domestic regulations should apply only to the good itself; regulations relating *only* to the process or production method (PPM) by which the good is produced should not be applied to imported goods *with substantially similar characteristics to the domestically produced good*.[15] The purpose is, 'to ensure that internal measures not be applied to imported or domestic products so as to afford protection to domestic production'.[16]

Whilst Article XI is the rule, there are exceptions, most of which are specified in Article XX. The USA argued, *inter alia*, that it might use Articles XX(b) or (g) of the GATT in its defence. Article XX(b) permits measures 'necessary to protect human, animal or plant life or health', and Article XX(g) permits measures 'relating to the conservation of exhaustible natural resources if such measures are made effective in conjunction with restrictions on domestic production or consumption'. But the header, or *chapeau*, of Article XX prohibits such measures if they are merely a disguised restriction on trade, which appears to be the case with the USA using the MMPA to prevent tuna imports.

The panel reported that neither Article XX(b) nor Article XX(g) was intended to be applied extrajurisdictionally.[17] Furthermore, it was pointed out that perfectly legitimate means exist to enable consumers to distinguish 'dolphin-friendly' tuna from dolphinicidal tuna, namely labels (in the USA the Flipper 'seal-of-approval' is one of these labels), which the panel ruled acceptable under the GATT.[18]

In 1992, then, it could be said that the judges on the GATT Dispute Panel felt that, at least with regard to dolphin conservation, the welfare of US inhabitants was better protected by ensuring that free trade prevails rather than by protecting US tuna fishermen from Mexican imports. This is, of course, consistent with the guiding principles of the GATT, whose objective is to provide a forum for enabling countries to overcome vested interests.

There is a direct parallel with the Common Law in England, which was developed from the twelfth century onwards through the King's Bench.[19] By establishing a common framework of rules, the law of the King's Bench enabled trade to take place more effectively because, whereas the local courts might favour powerful local parties, the king's court would be largely indifferent to local parties' interests (though of course the king would be subject to powerful national interests). In a similar way the GATT offers a forum in which to extend a system of impartial, open, clear rules to the whole world, at least where trade between nations is concerned.

Tuna–Dolphin II

Two years later, a second case, between the European Union and the USA, was brought before the GATT Dispute Panel (DP) with almost identical facts. In this case, the panel again reported that the USA could not use Articles XX(b) or (g) of the GATT in its defence. The DP noted that 'the tuna is embargoed simply on the basis of a country's policies, regardless of whether an individual tuna may have been caught without harming dolphins.'[20]

However, in this second case the DP accepted the general proposition that Article XX(g), which relates to the protection of exhaustible natural resources, could be applied extraterritorially under certain circumstances.[21] The fact that dolphins are not a globally endangered resource was considered to be significant (objections to dolphin by-catch seem to result from a poorly justified view of the intelligence of these cetaceans).[22]

Despite the fact that these reports were not adopted, the US made changes to its domestic legislation in broad compliance. Some environmentalists, who had lobbied for the MMPA and for its application to Mexican tuna, were irked by this and have sought to establish an MEA for dolphin protection.[23] They also cite the case as a reason to amend Article XX to allow for more general extraterritorial application of environmental laws. The effect of such an expansion would be to undermine the whole premise upon which the trading system is based, since it would effectively permit arbitrary restrictions on what can be traded.

Shrimp–Turtle

A similar dispute arose four years later in 1998 when India, Pakistan, Thailand and Malaysia found themselves on the wrong side of a US ban on imports of shrimp caught without the use of Turtle Excluder Devices (TEDs). The ban was made in reference to Section 609 of the Endangered Species Act (ESA), which lists turtles as an endangered species and empowers US regulators with the authority to impose measures to protect them. The initial ruling of the dispute panel was clear: the import ban was an unjustified restraint on trade (in breach of Article XI) and could not be justified under Article XX.[24]

However, the USA appealed the ruling. The appellate body (AB) of the WTO overruled the panel's decision, arguing that the panel's interpretation of Article XX was incorrect. The AB introduced a new two-stage test, which involves deciding whether the regulation meets the criteria of the appropriate clause in Article XX (b, d, g, etc.), and then deciding whether the criteria outlined in the *chapeau* of Article XX are met.[25]

The AB said that the US ban on shrimp imports passed the first part of the test. It deemed the measure consistent with GATT Article XX(g) on the grounds that turtles are listed as endangered under CITES and that the use of TEDs is reasonably related to the desired end of conserving sea turtles.

However, the AB deemed the ban to have failed the second part of the test because of the way it had been implemented, which meant that it was both arbitrary and an unjustifiable discrimination on trade between countries.[26]

All things considered, the AB's ruling amounts to a significant concession to the environmentalists, who had for years sought to impose Western environmental laws on poor countries by threatening those countries with trade sanctions. Moreover, the AB clearly spelled this out, towards the end of its opinion, stating:

> We have not decided that the protection and preservation of the environment is of no significance to the members of the WTO. Clearly it is. We have not decided that the sovereign nations that are members of the WTO cannot adopt effective measures to protect endangered species, such as sea turtles. Clearly, they can and should.[27]

Thus, if *Shrimp–Turtle* is followed, any country can impose restrictions on products on the basis of how they are produced or processed so long as the country imposing the restrictions is able to show that the measure is necessary to protect the environment and is carried out in a non-discriminatory manner.

After the AB ruling and after the USA complied with the requirement to impose the restrictions in a non-discriminatory manner, Malaysia launched a new challenge against the extraterritorial application of Section 609, but was again defeated.

What did this ruling mean to Asian shrimp fishermen, who might earn as little as US$500 a year? For such a fisherman, a Turtle Excluder Device costing US$150 is a significant cost. If he receives no economic return from using one, why should he do so? The fishermen continue to export shrimp to other countries who have not imposed such a ban, but receive less money for their product because purchasers do not face competition from the USA. Thus, surely it would be better to give the shrimp fishermen an economic incentive (for example in the form of a cuddly turtle-friendly symbol) to both procure new technologies and likewise generate more income.

Sea turtles are of some importance as an economic resource to some people in poor countries. But because environmentalists successfully lobbied for trade in sea turtles and products derived from them to be banned under CITES, their value has been significantly reduced.[28]

To answer Professor Pearce's second test (that production-related damage is a legitimate feature of the importing nations' loss of well-being), we can say that opinion at the Dispute Settlement Body of the GATT/WTO has shifted.

If this shift reflects a genuine change in public opinion, then it would be evidence that the importing nations' well-being is best served by permitting such restrictions on trade. However, I believe that the shift more accurately reflects a desire on the part of the members of the Appellate Body to avoid criticism from a small minority of anti-trade environmentalists. If that is true, it has not worked.

Immediately after the ruling, several environmentalists condemned the AB's decision, even though it represented a significant concession to their demands that trade sanctions be used to restrict imports on environmental grounds, and actually led to restrictions being imposed on imports of shrimp. Clearly the public posturing of environmentalist groups regarding the WTO has been more important than the actual decisions that have been taken. Ironically, even the people in turtle costumes on the streets of Seattle during the

WTO's 1999 ministerial meeting didn't have a clue about the implications of the decision. Nor did they know much about how the WTO works, or why there was a case. Mostly, they seem to believe that the WTO is part of a great global capitalist conspiracy.

For a court to have undermined the rules-based trading framework merely to appease a small minority of vocal activists is very troubling. The public still want high-quality produce at a low price, and mostly are not too concerned about how it is produced. Those who are troubled express their concern by purchasing produce labelled in ways that satisfy their preferences. Thus, consumers are harmed by bans on imports that are based on the way that goods are produced or processed, if the outlawed production or processing technique makes no substantial difference to the good itself. One would hope that *Shrimp–Turtle* was a minor aberration that will be corrected in the next case.

Implications of *Shrimp–Turtle* for Kyoto

Shrimp–Turtle appears to create a dangerous precedent. It says that trade restrictions may be imposed on any good if that good is produced in a way that is deemed to contravene the objective of an MEA.

In principle this means that parties who ratify an MEA could impose trade restrictions on other ratifying parties to the MEA, regardless of whether or not there are explicit trade sanctions in the MEA itself.

It is less clear as to whether *Shrimp–Turtle* would enable countries who are parties to an MEA to legitimately impose restrictions on non-ratifying countries. This depends largely on the extent to which an MEA is evidence of a commitment on the part of the international community to take action with regard to an ostensible environmental problem.

In the *Shrimp–Turtle* dispute, all the parties had ratified CITES (and, globally, 163 nations have ratified CITES). But the same may not be true for other potential disputes. Some environmentalists have argued that the EU should impose restrictions on imports from the USA on the basis that the USA has failed to implement Kyoto and that US industry therefore benefits from a subsidy by virtue of its lower-cost energy.[29] The EU has given tentative support to this proposal.[30] However, it seems likely that such restrictions would be WTO illegal (in addition to being harmful to both the citizens of the EU and the USA). At least until Kyoto comes into force, it would be very difficult to argue that the agreement represents the will of the international community.

The Kyoto Protocol, like other international treaties (including the WTO), was signed by the leaders of each country in their executive powers as the delegates of the people. However, in most states, such treaties are not directly binding on the people. This is for good reason: it forces the leaders of the signatory state to obtain the consent of the people they govern before binding them to commitments that might, on balance, be harmful. So, before Kyoto can come into force, it must be ratified by a sufficient number of member states, which in most cases must obtain consent from a broader representative body, such as Congress or Parliament.

Even if Kyoto comes into force, it remains unclear as to whether it can be used as a justification for imposing restrictions on trade with parties who have not ratified the Protocol. There are several reasons for this. First, it is not clear that even entry into force of the Kyoto Protocol represents the will of the international community. Certainly it represents the will of the ratifying parties, but are these parties representative of the international community? Although many are representative democracies, some are not, so the question remains: have these countries entered into the agreement freely and voluntarily and with the best interests of their people at heart, or were they cajoled or encouraged in some way? Some might even question the extent to which EU ratification is legitimate, given the dubious nature of the EU's democracy.

Second, if Kyoto is deemed to be the will of the international community, a question remains as to whether it is in fact the best solution to the problem of climate change, a question this book seeks to answer. If it is not, then trade restrictions imposed in the name of the Protocol would be illegitimate.

Third, if Kyoto is deemed to be the will of the international community and is deemed to be the best solution to the problem of climate change, a question arises as to whether trade sanctions are the best mechanism for enforcing the agreement. Given the analysis above, for most MEAs the answer would be no, so it is incumbent on those who demand the use of trade sanctions to explain why the case of Kyoto is peculiar. This point is addressed below.

EU protectionists

As if the *Shrimp–Turtle* decision is not bad enough, protectionist industries in Europe have been attempting to make formal changes to the GATT that would bias the system in favour of allowing WTO members to impose trade restrictions on environmental grounds.

Leading the charge has been the European Employers' Federation, the Union of Industrial and Employers' Confederations of Europe (UNICE), which represents big business in the EU. UNICE has suggested that Article XX should be amended to allow a rebuttable presumption that trade measures contained in MEAs are permissible and that 'MEAs reflect a broad consensus in the international community on how to solve global environmental issues.'[31] Such a change would not only reaffirm the *Shrimp–Turtle* decision, it would clarify its provisions regarding MEAs by affirming their status as justifying the restrictions, even against non-members.

That an industry body might take such a position seems at first quite astonishing. However, a moment's reflection enables one to see that for UNICE, this is a means of enabling the EU to impose restrictions on lower-cost imports from countries that are not compliant with various MEAs. It is blatant eco-protectionism.

If UNICE's suggestions were adopted, the presumption would be that trade restrictions based on non-product-related PPMs should be permissible generally, where they satisfy the objectives of an MEA, regardless of whether the state against which the sanctions are imposed has implemented the MEA or not. This would be an egregious abrogation of the rights of members of the WTO who have either not signed or not ratified MEAs.

In the context of Kyoto, a ratifying party such as the EU might legitimately impose trade restrictions on goods from a non-ratifying party (the USA or Australia, for example) on the basis that those goods benefited from lower-cost energy available in the USA or Australia because those countries have not ratified Kyoto. In fact, environmentalists have been calling upon the EU to impose such restrictions.

EU as protectionist

In 2001, the EU pressed for negotiations on the relationship between MEAs and the WTO. After resisting, WTO members finally caved in to negotiations on this matter after lengthy talks at Doha, Qatar, at the fourth ministerial meeting of the WTO. However, they circumscribed the negotiations to the relationship between MEAs and the WTO for parties to the MEA.

The EU has subsequently sought to interpret the meaning of MEA very broadly, so that an agreement of only three members might be deemed an MEA, and it has pushed for negotiations to be opened up so that the relationship between parties and non-parties could be codified.

The clear intention on the part of the EU is to enable the wide use of MEAs as a means of imposing restrictions on trade. It is almost without doubt that the intention of the EU is to impose restrictions on energy-intensive goods imported from the USA and thereby sate the desires of local protectionists and their environmentalist friends. (See Chapter 9, 'Bootleggers, Baptists and the global warming battle', for further discussion of the vested interests who have lobbied in favour of the Kyoto Protocol.)

It is clear that such protectionism, whilst beneficial in the short-term for a few industries, would be harmful in the long run to the majority of EU businesses and employees, as well as to EU citizens as a whole.

This brings us to Pearce's third test, which is to assess whether trade measures are the most cost-effective manner to bring about change in the product or process which causes the environmental problem. With Kyoto, would trade measures enable countries to deal with climate change?

The likely effect of such restrictions would be to slow economic growth in Europe and elsewhere. As a result, companies would have fewer resources available to invest in research and development into new production methods, which on average will be more environmentally benign, and many of which will be less carbon intensive. Moreover, by slowing economic growth globally, Kyoto prevents people from adapting to change, which is almost certainly the most cost-effective means of coping with climate change, at least in the short and medium term. In other words, trade restrictions based on Kyoto would make products and processes less, not more, sustainable.

It doesn't have to be like this

One might hope that *Shrimp–Turtle* was an aberration and that a future AB ruling would return us to the position we were in at the time of *Tuna–Dolphin I*, when non-product related PPMs could not be used as a justification for imposing trade restrictions.

However, another approach is to utilise the current negotiations in the WTO to clarify the position of MEAs vis-à-vis the WTO, reasserting the rule that non-product-related PPMs are not a legitimate justification for imposing restrictions on trade.

At Doha, Qatar, in November 2001, trade ministers from WTO members agreed, as part of a broader mandate, to negotiations on the relationship between the WTO and MEAs. Article 31 of that declaration says:

> With a view to enhancing the mutual supportiveness of trade and environment, we agree to negotiations, without prejudging their outcome, on:
>   (i) the relationship between existing WTO rules and specific trade obligations set out in multilateral environmental agreements (MEAs). The negotiations shall be limited in scope to the applicability of such existing WTO rules as among parties to the MEA in question. The negotiations shall not prejudice the WTO rights of any Member that is not a party to the MEA in question.[32]

Whilst this negotiating mandate is appropriately very narrow, it nevertheless offers an opportunity to clarify the relationship between MEAs and the WTO. Given the above discussion, it seems that the appropriate way to achieve this would be to declare 'a rebuttable presumption that trade measures in MEAs agreed prior to the entry into force of this amendment are not consistent with WTO rules.' On this basis, a country seeking to utilise such a trade measure would have to prove that the trade measure is consistent with GATT Article XX.[33]

In addition, it would be appropriate to declare, 'MEAs not ratified by all members of the WTO are not per se evidence of a commitment on the part of the international community to take action in respect of a particular environmental problem.' This will send a clear signal that the notion, introduced in *Shrimp–Turtle,* that MEAs provide evidence of such a commitment, was a mistake (perhaps based on the fact that

CITES, in spite of its faults, has been ratified by most countries in the world).

Conclusion

This chapter has demonstrated that trade measures are rarely, if ever a desirable, efficient or effective means of achieving environmental goals. Arguably the Kyoto Protocol is itself not an efficient or effective means of dealing with the problem of climate change. In light of these facts, it would be unwise to permit the use of measures to restrict trade in goods on the basis that they are produced in countries that have not ratified the Kyoto Protocol.

8 How Europe's risk regulations affect business

Martin Livermore

Introduction

In the last century, one of the most formidable challenges faced by humanity was the rise of socialist and communist states around the world. While that threat has for the most part subsided, the 21st century could be defined by a different kind of economic central planning, in the form of burdensome and excessive regulation of business and economic activity.

Judging by the policies it has pursued, the European Union seems to be the world's most risk-averse region. The EU's regulations are motivated by a distrust of business, the politics of special-interest groups, and the rise of the precautionary principle (and indeed, those factors are related). Such regulations have exacerbated the burden of government on commerce, businesses and consumers. We are heading towards what might best be called a 'disutilitarian' society – one which strives for the 'least harm to the greatest number' rather than the greatest good for the greatest number.

Business now faces an environment of uncertainty in Europe. In general, businesses are content to work within an established regulatory framework which defines a level playing field. Large businesses are generally in a good position to comply as they can cover the necessary overheads, and through lobbying, they have a hand at designing regulations with which they can comply.

However, excessive regulation benefits big business at the expense of small and medium-sized enterprises (SMEs). Two-thirds of European labourers are employed by SMEs, and this sector will provide much of Europe's future economic growth. However, it is at a huge disadvantage due to excessive regulation.

This chapter explores how Europe's policy towards climate change and global warming affects businesses. Europe's regulatory responses could have significant effects on a wide range of business sectors, and on economic development more broadly. Business's own reactions to climate change will prove important in balancing its core pursuit of pleasing consumers and delivering a profit to shareholders, while helping humanity to adjust to a changing environment.

Current regulations and their impact

Although much of the regulation relating to climate change exists in international agreements, it would be useful first to understand the general process of how regulation is introduced in the European Union.

The EU has increasing influence over a range of activities in its Member States. Although individual countries determine their own tax and overseas policies, for example, large amounts of regulation of other areas originate from Brussels.

There are two types of European legislation: regulations and directives. Once agreed, regulations have immediate effect throughout the Union after a defined grace period, and do not need to be passed separately into national law. Directives, on the other hand, place a responsibility on EU Member States to enact legislation within a defined time period, normally eighteen months. This has two main consequences: first, that detailed national legislation may differ to some extent between countries, and second, that it may come into effect at different times amongst Member States.

The introduction of EU legislation at the Brussels level is complex and somewhat opaque. This is largely because there are three primary European institutions involved in this process:

- **The Commission** is effectively the European civil service. It is comprised of fifteen separate Directorates General which cover agriculture, the environment, consumer protection and the internal market, amongst other issues. Officials draft regulations, but these individuals have no decision-making powers.
- **The Parliament** was once relatively weak, but now has a major influence on the final legislation which is passed. Committees

of Members of the European Parliament (MEPs) review and
propose changes to the Commission's draft legislation, and
there is then a first plenary vote. This legislative draft is passed
back to the Commission, which (along with the Council)
makes revisions to achieve a compromise before the process is
repeated and legislation is approved by a second parliamentary
plenary vote.

- **The Council** is the final decision-making body and its members
  are Member State politicians. The body has a number of
  subdivisions (for instance, the Agriculture Council, made up of
  Member State agriculture ministers) and it is the appropriate
  one of these which reviews and approves proposed legislation
  as it progresses. The Council must approve legislation after it
  has been passed by the Parliament.

Whichever country holds the European Presidency (currently rotated
between the fifteen Member States every six months, a system due to
change on enlargement of the EU in 2004) signs the legislation, which
is then published in the Official Journal and comes into effect.

In addition to the EU's own law-making, there are some areas of
EU policy set by international agreements, such as the Framework
Convention on Climate Change and the Kyoto Protocol, which effectively regulate other areas. Exactly how the EU views its obligations
under the FCCC and Kyoto, and the policy which flows from its view,
will be considered below. As implementation of the Kyoto agenda
gathers momentum worldwide, we can expect to see an increasing
range of regulations and directives introduced to give this the force of
European law.

Moving from the general to the particular, we should examine the
current situation in the climate change area.

Climate change and global warming

Climate changes: it always has and always will. However, there appears
to be a widespread consensus among many governments, international
agencies, some of the scientific establishment, and many NGOs, that
the earth's climate has entered a period of unprecedented warming due
to the release of increasing quantities of 'greenhouse gases', primarily
carbon dioxide through the burning of hydrocarbon fuels.

Consequent to this conclusion, international political deliberations led to the formulation of the Kyoto Protocol, where signatory parties from the developed world have committed to significant reductions in their outputs of greenhouse gases. The initial period covered by Kyoto is 2008–12, with an overall target of a reduction in CO_2 output by 5% over 1990 levels. However, the intention is that Kyoto is only a first step towards more radical emissions reductions.

The Kyoto Protocol is structured such that industrialised countries – calculated to be responsible for 55% of the world's current emissions of carbon dioxide – are listed in Annex 1. To come into force, the Protocol must be ratified by 55 of the total signatories to the Convention and by Annex 1 countries responsible for at least 55% of total emissions in the baseline year of 1990. The European Union Member States have enthusiastically signed up and have agreed an overall regional target of 8% by 2012. The USA has been hesitant towards the process and the Protocol, stating concerns over the Protocol's effects on the economy, and it has clearly stated that it has no intention of ratifying. The USA accounts for approximately one-quarter of global CO_2 emissions, so this rejection is significant.

Many parties including the EU have set emissions targets to comply with the provisions of Kyoto, and have encouraged Russia to ratify Kyoto so that the agreement enters into force worldwide. The EU published a White Paper which established targets for energy use and generation from a range of sources.[1] A subsequent White Paper set out more specific plans for renewable energy.[2] While markets for hydrocarbon fuels have been deregulated to the benefit of consumers, the EU has started the process of regulating future energy supplies to reduce emissions and to increase the contribution of renewable energy sources. This process will have a minimal effect on global warming, as Bjørn Lomborg points out:[3] 'Several models have calculated that the consequences of Kyoto will be a temperature increase by 2100 of around 0.15°C less than if nothing had been done.'[4] Yet the Protocol will be a significant impediment to economic growth.[5]

Commissioner Wallstrom commented on the 2003 assessment by the European Environment Agency (EEA) of greenhouse gas emissions:

> The European Union is moving further away from meeting its commitment to achieve a substantial emissions cut under the

Kyoto Protocol. The progress we have made already needs to be backed up by additional measures. Especially the Member States that are not on track in reaching their targets urgently need to take additional action.[6]

This warning of the difficulties of achieving the emission control targets is borne out by the EEA figures of actual progress towards reaching the 2012 goals.[7] The EEA reports that total GHG emissions in the EU were reduced by 3.5% between 1990 and 2000. However, further analysis reveals a picture which is more complex:

- Half the total emissions reductions occurred because of special circumstances in two Member States. The deregulation of the UK power generation industry led to a major switch from coal- and oil- to gas-fired power stations, and specific measures were put in place to reduce nitrous oxide emissions in the chemicals sector. In Germany, changes were brought about by restructuring in the eastern Lande and increased efficiency in power generation.
- In the last two years of the decade, carbon dioxide emissions actually rose.
- Nine out of fifteen Member States were not on track to meet their burden-sharing targets.
- Reductions in GHG emissions from most sectors, including industry, were partially offset by a 20% increase in emissions from transport.

The assessment of the ten EU accession countries shows an overall significant decrease in emissions during the decade. This is hardly surprising considering the dire state of their heavy enterprises, which were formerly governed by the public sector, and the massive restructuring which occurred in those economies during the decade.

The EU and Member State governments face the choice either of missing the targets or forcing economic changes which, at worst, could have a dangerously destabilising effect on society.

To some extent, obeying the strictures of the Kyoto Protocol fits with certain Member State energy policies. The contribution of North Sea oil and gas is decreasing: the UK, for example, will soon become a net importer of natural gas. As a region, the EU relies heavily on oil imports from politically unstable areas. A move towards self-generation of power from renewable sources makes a great deal of sense in this context, and the 'global warming' dogma makes the costs somewhat more acceptable politically. In a similar way, France is in the unique position of generating a high proportion of its energy from nuclear fission and thus it already has a low emission economy. In these circumstances, signing up for the Kyoto Protocol is politically popular and appears to have little or no downside.

The effects of regulation on businesses

Regulation affects business in many ways, both direct and indirect. The most obvious effect is to make business more expensive and difficult.

Energy is a classic area where costs are manipulated in pursuit of government policy. For example, in the transport sector the majority of the price which consumers pay for fuel goes directly to government in the form of tax. This is a marvellous way of generating revenue, as traffic increases inexorably. At the same time, governments can justify tax and duty increases in the name of environmental protection, as a way of limiting road use.

In fact, this cynical argument holds no water: road transport use in particular seems to be rather price insensitive. On the other hand, such increases will eventually meet resistance: the fuel price protests in the UK in 2001 effectively capped further government fuel tax rises. The protests were led by the commercial sector: UK lorry drivers were being put at a significant competitive disadvantage against other European competitors in what has become largely a European rather than a national market.

This particular example in the transport sector is an extreme case of a specific interest group influencing policy directly and very publicly. More generally, there is sustained lobbying activity behind the scenes by a range of pressure groups, each of which is seeking to maximise its own benefits. (See 'Bootleggers, Baptists and the global warming battle', Chapter 9).

The fact that governments are demonstrably influenced by lobbying has led to a mushrooming of pressure groups and business lobbyists in the last few decades, each arguing their often narrow specialist agendas. On the issue of climate change and energy policy, the balance in Europe tends to be clearly in favour of the environmental lobby, as well as scientists (who obtain research euros from government), international agencies such as the United Nations, and the news media.

Business has almost entirely stopped opposing the lobbying of environmental groups, and instead is conspiring to ensure that regulations protect its own narrow interests. A particularly striking example of this was the dissolution of the Global Climate Coalition, following the initial defection of BP. The company took the path of least resistance by ensuring that it influences energy policy from the inside and is in a better position to benefit from it.

In the wider energy use area, the net effect until recently has been for deregulation of power supplies to reduce consumer prices in both the private and commercial sectors. However, the trend will now be for energy costs to increase, not for market reasons but because governments have started to impose measures to reduce energy use and carbon dioxide emissions to achieve the Kyoto targets. Ian Russell, Chief Executive of Scottish Power, recently estimated that by 2010, electricity prices will increase by 15%.[8] He suggested that regulatory pressures 'could add approximately 15% in real terms to unit costs by the end of the decade ... It could be more than this depending on circumstances, but the component parts are already clear. The energy efficiency commitment, carbon trading and the renewable obligations are guaranteed to put upward pressure on unit prices.' Would similar changes across the EU drive economies back into an inflationary spiral?

Estimates have been made of the effect of compliance with Kyoto targets on the economies of EU countries. For example, DRI–WEFA produced a series of reports in 2002 for the International Council for Capital Formation, looking at two cases for CO_2 emissions.[9]

- The agreed first period (2008–12) commitment plus reduction by 60% from 1990 levels by 2050.
- The agreed first period commitment plus zero net emissions by 2050.

Some of the main findings for key countries up to 2020 are given below. For the first period, the EU has a target of 8% CO_2 reduction, but individual country targets vary widely.

- In the **UK**, emissions need to be reduced by 11% by 2010 and 14% by 2020. The result for 2010 would be a 57% increase in the price of natural gas and 59% increase in the cost of electricity. The net effect on the economy would be significant: a projection of an increase of 400,000 people unemployed by 2010 and a loss of 1.9% in GDP for the same year. Going forward to 2020, unemployment would continue at the same level, and GDP would still be 1.7% below baseline projections.
- **Germany** has similar targets: 10% by 2010 and 14% by 2020. Industrial energy prices would increase, with natural gas rising by 27% and electricity by 60%. Overall effects on the economy would be greater. GDP would fall 2.7% below the baseline for the first period, and would continue at 2.5% below by 2020. Unemployment would increase by about 1 million in 2008–12, reducing by only about 20% of this by 2020.
- The effects on **Spain** would be dramatic, since emission reductions of 25% and 27% would be needed by 2010 and 2020 respectively. This would lead to industrial gas and electricity prices increasing by 63% and 70%, and petrol by 18%. The GDP would shrink by 4.8% and by 2010, unemployment would rise by 850,000.

There will clearly be a very significant effect on the EU economy if these reductions go ahead. But few countries seem to be on course to meet their initial targets, and it is unclear how much political will exists to force the structural changes needed. Perhaps voluntary free market initiatives will have a larger part to play than has previously seemed likely.

In a world where the Kyoto Protocol and subsequent treaties are fully implemented, globalisation and trade will mean that business will move to wherever costs are lowest. This is a trend that we have seen for decades, as heavy industry moved to East Asia, assembly plants proliferated in countries such as China and Mexico, and even call centres began to relocate to India. In the future, manufacturing capacity could increasingly move from high-cost to low-cost energy countries.

Not that the picture is necessarily equally gloomy for all businesses in Europe; there are always those who benefit. These we might term the 'silver lining' companies. These would include specialist firms offering their expertise in increasing energy efficiency and reducing emissions. Waste management and environmental protection is a growing business; such companies would undoubtedly focus some of their efforts on activities which reduce overall emissions of GHGs.

How has business responded to climate change?

In 1989, at the very beginning of the global debate about climate change and global warming, a number of businesses formed the Global Climate Coalition to lobby against policies that were being developed at the international level to address global warming (including the Kyoto Protocol). The GCC was obviously a self-interested effort – the members were concerned that governments might enact policies that were short-sighted, in light of scientific uncertainties, and could be costly for their own operations as well as for consumers. However, they did succeed in influencing the debate to consider the economic costs of policy actions, which might have otherwise been missing.

The GCC's members had divisions, though, and they also faced the constant ire and criticism of their environmental critics. In 1993–94, splits emerged in the GCC on issues such as carbon trading and voluntary emissions restrictions, and the GCC lost some of its lobbying power.

In the meantime, European governments were working overtime to negotiate the Kyoto Protocol, and European businesses were being pressured to agree to restrictions on emissions. In some ways, their efforts reflected the European Commission's need to find a justification for its existence and to raise revenues. Vested interests, mainly from the environmentalist community, provided the impetus for the EC to push forward on Kyoto.

In 1997 there was strong international pressure to finish the agreement, and to set targets and timetables. At the time, European businesses were hard-pressed to resist the regulators and it was increasingly clear that they needed to make a tactical decision. Thus, two major businesses decided that they could not continue resisting the basic principles of Kyoto – BP/Amoco withdrew from the GCC, and shortly after the Brent Spar incident Shell also withdrew. These defections led to the withdrawal of other companies, and the ultimate collapse of the GCC. Of course, when BP/Amoco and Shell accepted the fundamental premise of climate change, it allowed them to influence the regulations that they would ultimately have to bear, both at the European level and at the international level.

Business's approach to climate change in recent years has been somewhat divided between businesses such as Shell and BP, who are attempting to shed the image of being 'big, polluting, profit-mongering oil companies' by publicising their non-oil based activities, and those such as Exxon, Toyota, General Electric and Schlumberger, in conjunction with Stanford University, which have joined together for a 'blue-skies' research project called the Global Climate and Energy Project (G-CEP).

'BP' – a highly recognised brand and trading name – originally stood for 'British Petroleum'. Now, with a new (green) logo, BP portrays itself as being 'Beyond Petroleum' which has afforded some beneficial short-term publicity. BP is a 'leading photovoltaic power supplier', but this is merely reported as part of an overall 'gas, power and renewables' sector, in turn accounting for only 3.5% of group operating profit. In review of its achievements for 2002, the company reports that it cut its emissions of greenhouse gases by 10% and committed to maintaining the same net level of emissions for the next ten years. For this major energy company, the view seems to be that freedom to operate is more easily guaranteed by subscribing to the dominant view on climate change than opposing it on rational scientific grounds. But it remains to be seen whether 'beyond petroleum' will prove a viable, long-run strategy for a company whose profits are still almost entirely based on hydrocarbon energy.

The Ford Motor Company has faced the same problems as General Motors, Daimler-Chrysler and other major car manufacturers: overcapacity and increasingly stringent regulation. Despite producing ever safer and more economical vehicles, they face constant pressure from environmental groups, who often has a major influence on government policy. Ford has responded by working to establish its green credentials while concentrating on its mainstream business of making and selling cars profitably. For some time, Ford has produced cars which run on LPG and ethanol, and cars powered by fuel cells will be introduced in 2004. On the environmental front, Ford has pursued a strategy which probably makes sense – decreasing its own energy use, and increasing the fuel efficiency of its vehicles.

Shell has similarly called for the world 'to shift to low or zero-carbon alternatives such as solar, bio-fuels and fuel cells running on hydrogen.'[10] It has publicly advocated its investments in renewable fuels, and has criticised its competitors, which have not lobbied in favour of the Kyoto Protocol. In 2001, it announced that it would spend £714 million on research into renewable energy sources.

The G-CEP project illustrates a very different approach by companies to the issue of energy sustainability. According to its website, the purpose of G-CEP is 'to unite scientific researchers and private industry from around the world in the search for commercially viable technologies that foster the development of a global energy system where greenhouse emissions are much lower than today'.[11]

Their justification for the project is the 'possibility that the world will need an energy system that has much lower emissions of CO_2 and other greenhouse materials to the atmosphere in the future'. The businesses who support G-CEP have also been subject to extreme criticism by environmental campaigners, but feel that the best path forward is one which recognises that the fundamental purpose of business is to provide goods and services to consumers, and to generate revenue for shareholders, but still be mindful of the future. It is likely that such a strategy will pay off in the future – for the earth (if lower emissions of CO_2 are necessary), for shareholders and for consumers.

'Greenwashing' and its progeny

The 2002 World Summit on Sustainable Development in Johannesburg provided an opportunity for strange bedfellows to publicise their relationships. At the WSSD, the World Business Council on Sustainable Development and Greenpeace co-sponsored activities to encourage governments to ratify the Kyoto Protocol. In a joint statement, they said:

> Given the seriousness of the risks of climate change, and the need
> to reduce greenhouse gas emissions, [Greenpeace and the WBCSD]
> are shelving our differences on other issues on this occasion and
> call upon governments to be responsible and to build the
> international framework to tackle climate change on the basis of
> the UN Framework Convention on Climate Change and its Kyoto
> Protocol. We both agree that this is the essential first step.[12]

But despite the efforts of Shell and BP to be green and progressive, environmental campaigners are still displeased because most of their revenue continues to come from hydrocarbon fuels. An NGO awarded both Shell and BP (two major companies who form part of the WBCSD) a 'Greenwash' award during the World Summit.

Environmental campaigners have cynically responded to business's efforts to promote less carbon-intensive energy by referring to those efforts as 'greenwashing.' They claim that business is 'greenwashing' when it engages in any economic activity which is not its core activity:

> if a company's core (or main) business is based primarily on an activity which has been identified as significantly contributing to environmental pollution or destruction, there is a strong presumption that any assertions that it supports environmentally sustainable development are greenwash.[13]
>
> In our experience, for every company that is genuinely trying to improve its social and environmental performance, there are many more that are merely engaged in 'greenwash' (PR exercises designed to give the impression of a greener company) but with little real change to corporate activities.[14]

Of course, companies' efforts are motivated by enlightened self-interest, since they may well be investing now in technologies which will provide major new revenue streams for them in future. In other cases, business reacts to pressure groups to address particular claims, in order to secure freedom to operate in their mainstream businesses. But either way, business is responding rationally to concerns driven by environmental campaigners.

Now it seems that such accusations have led environmental campaigners to suggest that a global agreement on corporate accountability should be negotiated and implemented, so that corporations will be subject to an international body of law which governs their activities.

Friends of the Earth International (FoEI) lobbied for such a convention at the WSSD:

> Unless all corporations are made equally accountable for their environmental and social impacts there remains little incentive for a general improvement in behaviour. What is more, those

corporations which want to become more socially responsible are being held back by competitors who can undercut them by continuing to externalise costs and by demonstrating no responsibility.[15]

Such an agreement would entail a global set of 'binding corporate accountability' rules to make corporations legally accountable not just to their shareholders but to any group who claims that they have been harmed by a business's activities. FoEI and other organisations continue to lobby, both explicitly (ie. for a specific UN convention) and implicitly (by including such rules in other agreements or through the WTO), for this agreement.

Of course, most smaller businesses are focused on day-to-day activities, and they are less subject to outside criticism from NGOs and the public. They also are less involved in lobbying to design regulation to benefit their interests. In the future, small businesses will probably face a regulatory requirement to address corporate social responsibility issues, and this will mean increasing costs of compliance. In practice, many businesses already cover these issues in a piecemeal fashion via employment law, regulations on waste management and initiatives designed to ensure the meeting of Kyoto targets, such as the UK climate change levy. Yet again, it will be the innovative small company sector which will suffer disproportionately from increased regulation.

Business and the environment

But are global rules needed to force businesses to be environmentally sustainable? There is good reason to believe that companies which operate in a market framework have largely always acted to reduce industrial waste and pollution, thus benefiting the environment. Pierre Desrochers, an economist who works on industrial ecology issues, explains

> Virtually all contemporary experts on sustainability assume that traditional economic development was characterised by a linear approach in which materials and energy were extracted, processed, used and dumped in a linear flow into, through and out of the economy. Much historical evidence, however, indicates that industrial resource recovery was much more widespread than is currently thought ...[16]

Today's companies are getting better at explaining their core activities in such a way that consumers, and the public, understand the environmental benefits of their production processes and products. This is a challenge for businesses, because, as Desrochers suggests, 'Individuals are more familiar with municipal waste disposal practices than industrial behaviour towards by-products ... Current environmental regulations are squarely based on the notion that industrial by-products are a nuisance to be destroyed rather than potentially valuable inputs.'

Desrochers concludes that:

> Our ancestors did not expand their economies by simply doing more of what they had already been doing, but by inventing new kinds of goods and services and by creating wealth out of what had hitherto been considered valueless things. It therefore seems fair to say that all of today's recyclable products were considered waste at one point in time, before value was created out of them through the use of human creativity and entrepreneurship. The market process is, of course, not perfect ... It may be that in today's economies, regulatory barriers and price-distorting subsidies are more serious obstacles to creating value out of by-products than traditional market incentives.[17]

Successful companies today are no different from those in the past. Their business is to satisfy consumers and generate profits for shareholders, and often to reinvest their profits into researching and developing technologies which yield less energy- and resource-intensive production of goods and services. Businesses, of course, do not set out with the intention to pollute the environment, use up natural resources, or encourage profligate consumption, as alleged by many environmental campaigners. As Desrochers indicates, businesses naturally strive for improved efficiency and profitability – and generally, this means that their activities, and the goods and services they produce, are more and more environmentally benign. What has happened, though, is that businesses are better at communicating to consumers and the public about how their activities fit in with, and most importantly do not undermine, people's core values and priorities.

In either case, it is clear that some environmental NGOs are motivated by a desire to both lighten humanity's footprint on the earth and promote widespread human well-being. Moreover, some of these groups even understand that it is important for business to operate in a flexible and conducive environment to achieve such goals. However, other environmental NGOs, even some considered part of the 'establishment', clearly seek to undermine business, or at least they are uninterested in the impact that regulations might have on business, because they are narrowly focused on environmental protection, regardless of the benefits or problems for humanity which result from it.

It is these NGOs who are largely cynical about business and the environment. Their self-interest is obviously not to earn a profit for shareholders and to please consumers, but to convince people of the urgency of their claims. To that end, they will continue to promote the myth that the earth is in peril because of consumer culture, and that business is the cause.

Alternative approaches to risk regulation

The primary purpose of good regulations should always be to protect the citizen or society at large. The trouble is that the net of protection is cast wider and wider in Europe especially, with no well-defined limits to government's intervention in how companies conduct their business. As we have seen, continued reliance on the precautionary principle means that there is no accounting for cost/benefit relations, and this is not a rational basis for regulation.

Since individual regulations have specific aims, it should be possible both to justify their introduction and to monitor their effect. The regulatory cycle, in an ideal world, would therefore look something like this:

- Identify preventable harm and/or desired positive outcome.
- Propose and assess possible regulatory routes to achieve the outcome, including full cost/benefit analysis and assessment against agreed objective criteria.
- If justified by the analysis, introduce appropriate regulations with effective enforcement regime.
- Ensure that regulation is broadly based. Consider input from all stakeholders but do not allow one to have more influence than another.

- Monitor the effect of the regulation, both for the desired outcome and unexpected consequences.
- Review the data after an agreed time period.
- If necessary, repeal or modify the regulation.

In practice, regulation is too often influenced primarily by lobbying from one group of stakeholders – whether business, environmental groups or other special interests – or it is introduced with insufficient evaluation of alternative approaches. Even worse, if such regulations prove ineffective, they are often either replaced by more stringent versions, or simply left on the statute books to be enforced arbitrarily.

Lobbying is, unfortunately, an integral part of the development of regulation. Sometimes, its effect can be pernicious, particularly if one view dominates. This is clear from the increased influence of the environmentalist lobby, particularly in Europe's enthusiasm for meeting Kyoto targets, and from the influence of large businesses which have a hand in designing regulations which are designed to help their bottom line, but – whether intentional or not – provide a competitive disadvantage to their smaller competitors.

Climate change policy will have a great and widespread effect on both business and society. Are there more constructive ways in which this issue could be tackled? Indeed, are there bodies other than national governments who are better placed to take overall responsibility? Politics may be the art of the possible, but it also has short time horizons. Political bodies which look no further into the future than the next election are arguably ill-equipped to handle the development of long-term, consistent strategies to deal with issues of global significance on a generational timescale.

Business, unfortunately, often has equally limited time horizons. The need to meet annual (and quarterly) financial targets often leaves long-term strategy by the wayside. Activists in the environmental and consumerist organisations, on the other hand, often have a much clearer, if blinkered, vision of the future: they know what their long-term goals are, and use whatever tactics are necessary to work towards them. Clearly, short-term successes are needed, but the effectiveness of their lobbying is derived from the tenacity that comes from moral certitude.

But the over-regulation we are now suffering is not just due to lobbying: it is also cultural. Our current manner of assessing and dealing with risk has two major drawbacks. First, we have excessively pre-cautionary regulation, and second, we have excessively precautionary enforcement of otherwise workable regulation.

The answer to these problems could lie in depoliticising the en-forcement process and entrusting risk assessment and management to experts – bodies of independent civil servants who are charged with rational decision-making based on sound scientific advice. Commit-tees of experts from different countries who assess the same dossiers may not always agree fully on the amount of data required, but align-ment on the scientific facts is common. It is only when the recommen-dations are passed to politicians that irrational decisions are sometimes made.

All the evidence is that Americans are equally well protected from risk as their European cousins: rational decision making can clearly be effective. However, this is not the full story: US regulations are still made politically and influenced by lobbying, which tends to be in favour of business interests. The US economy and business sector is not disadvantaged by this overall, although the balance favours major business interests and this may not be beneficial for smaller compa-nies. The conclusion is that regulation, however intended, will be a disadvantage to smaller businesses, while favouring larger companies.

Deregulation of risk management

The problem with creating a global framework and targets to address climate change is the same as that which beset the Communist bloc and led to its downfall: central planning and outcome-based ap-proaches do not work. The solutions envisaged by the Kyoto Protocol and beyond comprise enormously complex and interdependent sys-tems to force down emissions of greenhouse gases. Such systems have too much inertia to react quickly to changing circumstances, and are unlikely to achieve anything.

So rather than relying exclusively on government to regulate risk, what framework would encourage businesses to address risks while still maintaining profitability?

Our goal should be to enable ourselves to react to new knowledge and new circumstances. So far, flexible market economies, with regu-lations based on sound science and rational risk assessment, and a competitive and fair playing field for all actors, have been shown to be the best system to both generate wealth and protect the environment. In the decades after World War II, it was the free market economies of western Europe which became increasingly prosperous while pollution levels were dramatically reduced. By contrast, the old eastern bloc countries had stagnant economies with appalling levels of pollution.

Many existing factors will continue to make industry more energy-efficient, if only through the competitive impetuses of the market. Equally, there are many good reasons to rethink and improve transport systems to benefit everyone, rather than rely on punitive regulatory measures to reduce CO_2 emissions.

Businesses, particularly small ones, would benefit from this system, because they would be free to do what they do best: provide their customers with products and services they want, at prices they can afford, and to do so profitably. If customers or citizens are dissatisfied with their practices, they can change the company's behaviour by not purchasing their products, that is, by boycotting them.

We are uncertain what will happen to the earth's climate. In fifty years' time, we may or may not be worried about global warming; we may once again be concerned about global cooling. However, it is certain that scientific developments in the energy and transport sectors will have been enormous. In turn, business's application of this science will bring huge benefits to society. These benefits will inevitably present some negative effects. But the appropriate response is to rationally address and manage such problems, rather than attempt to eliminate risk through regulation, or even worse, through prohibiting such innovations altogether.

Conclusions

Present EU energy policy is leading us into possibly the worst of all possible worlds: a strategy which is unnecessary, because the science of climate change is uncertain; ineffective, because the EU's reductions in emissions will not significantly affect climate change; and negative for nearly everyone in society.

Business will carry a large proportion of the burden – their profitability and overall economic growth will suffer. Consumers will suffer because the price of goods will increase, since business will pass on higher costs of compliance with regulation to consumers.

Larger businesses, particularly those able to influence regulation to their own advantage, may increase their competitiveness, but small businesses are likely to be most negatively impacted. Those businesses which have the flexibility to do so will increasingly relocate to developing countries and contribute to the decline of Europe as an economic force.

However, the internal divisions of the European Union make it difficult to pursue policies to which major Member States are not committed. Even then, ensuring compliance is difficult, and few effective sanctions are available against those who fall out of line. As businesses will ultimately pass on the costs of the Kyoto Protocol to consumers, will European citizens tolerate its impact on their lives? If not, public opinion may force a change in policy to allow us to adapt to change.

Companies have already helped us to solve a huge problem during the twentieth century, which was feeding a growing population while maintaining the earth's wild and biodiverse land. Modern agricultural technologies developed and marketed by companies have allowed us to largely achieve that goal, though progress must still be made.

Today, the world's most urgent environmental problems are local – indoor air pollution, dirty water and lack of sanitation, and preventable diseases are the cause of the most problems in the world, and these are found mostly in poor countries. Companies should play an important role in solving these problems.

Global warming may or may not be a reality, but business should be enabled to do what it does best: innovate and produce goods and services in a manner which ensures that humanity's footprint on the earth is ever-lighter. And companies should realise that consumers are also citizens – they care about protecting the environment, and want to feel secure that the activities of business are not undermining that goal. Whether or not global warming presents a real threat, business is an agent of change, and will continue to help humanity to adapt to change.

Part Four

The Broader Context of Climate Change Policy

9 Bootleggers, Baptists and the global warming battle

Dr Bruce Yandle and Stuart Buck

Introduction

Since the 1997 International Conference on Climate Change in Kyoto, Japan, the world's industrialised nations have been grappling with negotiations over the Kyoto Protocol. Passionate expressions of concern about global warming have given way to tough political bargaining over who bears the pain, most recently seen in the tough negotiations at the Sixth Session of the Conference of the Parties to the Protocol (also known as COP6) in November, 2000.[1]

The most stunning blow to the prospects for the Kyoto Protocol came in March 2001, when the current US presidential administration decided to reject it.[2] Though some congressional members have publicly groused about the decision, there is little doubt that most of Congress supports Bush's action.[3]

The international response to Bush's action? Generally speaking, outrage. European leaders accused Bush of being 'irresponsible' and 'arrogant', while Margot Wallstrom, the Environment Commissioner for the European Union, said, 'We cannot allow one country to kill the process.'[4] British Environment Secretary Michael Meacher said that Bush's decision could make the planet 'uninhabitable',[5] and Deputy Prime Minister John Prescott accused the United States of 'free-riding' and of sitting in 'glorious isolation'.[6]

The outrage expressed by the Europeans may, of course, be slightly hypocritical. None of the EU countries had ratified the Kyoto Protocol themselves when the USA pulled out. And little wonder: in the words of *The Economist*, 'The EU's dirty little secret is that very few of its own members are on track to meet their tough Kyoto targets by the deadline anyway.'[7] Canada's Environment Minister, David Anderson, has even said that the EU deliberately sabotaged the Kyoto negotiations by 'adopt[ing] a position they knew would force the United States to pull out'.[8] Why? In the words of journalist Gregg Easterbrook, 'Because Europe didn't want to do anything about the greenhouse effect but wanted the United States to take the blame.'[9]

The fury over Kyoto's failure is misplaced. Good evidence suggests that the Kyoto Protocol would have been a potentially huge drag on the US economy, while producing environmental benefits that would in all likelihood be too small to measure. Moreover, the burden borne under Kyoto would have been unevenly placed: if the US had signed the Kyoto Protocol, it would have been required to reduce the emissions projected for the years 2008 to 2012 by some 40% to meet its Kyoto goal.[10] Yet the developing world, which emits large quantities of greenhouse gases (GHGs), would have faced no limits at all.

With the majority of the world's population free to expand emissions, it is highly likely that total carbon emissions would have continued to increase almost unchecked by Kyoto, rather than decreasing. This fact suggests that Kyoto negotiations involved much more than a public-spirited commitment to reduce carbon emissions. Far from an objective solution to climate change, the Kyoto Protocol created a new arena for nations, groups and companies to pursue their special interests.

This chapter explores the 'bootleggers and Baptists' theory of economic regulation to shed light on the Kyoto negotiations.[11] This theory best explains the actions and manoeuvring displayed by many countries and corporations, in both their support for and opposition to Kyoto. We conclude that various nations and corporations have tried to influence Kyoto's terms to serve their own parochial interests.

The Kyoto Protocol

The origin and terms of Kyoto

The fundamental premise of the Kyoto Protocol is that developed countries, which are large energy users and GHG producers, should bear the brunt of reducing emissions to avoid climate change. By the years 2008 to 2012, developed countries are required to bring their total GHG emissions to 5% less than they were in 1990.[12] The USA in particular was expected to reduce emissions to 7% below 1990 levels by the years 2008 to 2012.[13] This decrease in emissions is to be achieved by reducing GHG emissions and by creating 'sinks', that is, by increasing absorption of GHGs by forestation or other measures.[14]

Meanwhile, countries described as in 'the process of transition to a market economy' are treated with much more solicitude. These developing countries, which included China, India, South Korea, Mexico and some 130 other countries, have the option of using a base year period other than 1990 from which to measure their reductions (if any),[15] and will be given a 'certain degree of flexibility' by the Conference of Parties to the Protocol.[16] The result, as everyone admits, is that 'developing' countries will face no limits at all under Kyoto.

The Protocol allows limited trading in emissions. A given country can buy the right to emit a certain amount of carbon dioxide (CO_2) from another country that values the money more than the right to emit.[17] But such trading can occur only under certain conditions: the trading must involve a 'reduction in emissions by sources, or an enhancement of removals by sinks, that is *additional* to any that would otherwise occur';[18] the country buying emissions rights must be in compliance with its obligations under Articles 5 and 7 for measuring emissions;[19] and any trading must be 'supplemental to domestic actions' for the purpose of reducing emissions.[20] The requirement of supplementarity would have the effect of reducing the gains of trade on all sides.[21]

The Protocol also allows countries to band together in voluntary associations in order to have their emissions considered collectively[22] – a 'bubble' scenario that is widely understood as applying primarily, if not solely, to the European Union.[23] If the EU's emissions are considered as a whole, individual EU countries have a great deal more flexibility in meeting any individual emissions targets, without necessarily having to explicitly 'trade' emissions rights. If Greece has higher emissions one year for whatever reason, while another EU country has lower emissions, no trading need occur unless the overall sum of EU emissions surpasses the overall limit.

Two other features of Kyoto involve international cooperation on carbon reduction projects. Under 'Joint Implementation', two industrialised countries (or 'Annex I' countries) can create a joint project to reduce emissions in one country, with the reduction counting toward both their targets.[24] Specifically, any Annex I party can acquire or transfer 'emission reduction units', which result from projects that reduce emissions or enhance removals by sinks of greenhouse gases.'[25] A number of restrictive criteria apply to such projects.

The 'Clean Development Mechanism' feature is similar, except that it involves industrialised countries funding 'certified project activities' that would reduce emissions by *developing* countries.[26] Such projects would be allowed only if they produced 'real, measurable, and long-term benefits', as well as emission reductions that are 'additional to any that would' otherwise occur.[27]

What are the potential benefits of Kyoto? Possibly few. Because Kyoto requires only that industrialised countries cut back on emissions, it is highly unlikely that the goal of limiting total worldwide emissions to 1990 levels would possibly be met.[28] Even if the industrialised world manages to cut back to 1990 levels, the developing world's emissions would in all likelihood outweigh any reductions.[29] Predicted rates in CO_2 emissions between the years 1999 and 2020 estimate that the industrialised world will increase emissions at an annual rate of 1.2%, but the developing world will increase emissions at 3.7%.[30]

As one scientist put it, Kyoto 'won't prevent total greenhouse emissions from rising' because the 'cuts will be swamped early in the next century by increases in emissions from developing nations such as China and India ... '[31] Thus, British Environment Minister Michael Meacher has said that to achieve any meaningful climate control, industrialised countries would probably have to cut emissions by about fifteen times what the Kyoto Protocol would require.[32]

As a result, even scientists who support Kyoto often admit that it would make little difference to the earth's climate.[33] And yet another reason that Kyoto might not make that much difference is that many large corporations are already trying to limit energy use on their own.[34] After all, any reduction in energy use that can be made cost-effectively will translate into lower prices and higher sales for their products.[35]

The obvious question is, therefore, why was Kyoto supported so strongly, and why was the USA's rejection so vehemently condemned? Environmental groups still support Kyoto on the grounds that any carbon emissions reductions are worth achieving and overall emissions may decline eventually. But others, including companies and countries who want more market power, appear to have been more interested in the strategic possibilities offered by regulation under the

Kyoto Protocol. The economic distortion caused by Kyoto would have offered many opportunities for rent-seeking. The real reason for Kyoto's support is found in the 'bootleggers and Baptists' theory of regulation.

Bootleggers and Baptists

The central problem with global environmental regulation is that it is a classic public good.[36] To the extent that reducing greenhouse emissions is beneficial to the earth's climate, everyone will benefit, even those who refuse to contribute to greenhouse reductions. Because of these two factors, the relevant actors (corporations and countries) have an incentive to limit their own contribution towards any global environmental goal (i.e., to free-ride), while attempting to maximise the contribution of other actors.

Economic theories about regulation, however, are not necessarily complete. Not all small, well-organised special-interest groups will be able to see their regulatory goals put into action. The theory of bootleggers and Baptists,[37] a subset of the economic theory of regulation, further helps explain environmental regulation like the Kyoto Protocol. While powerful interest groups still matter, this theory suggests that efforts to achieve any given regulation will be most successful if at least two quite different interest groups work in the same direction – 'bootleggers' and 'Baptists'.

The term originates in the southern USA, where in the past and even today Sunday closing laws prevent the legal sale of alcoholic beverages. This is advantageous to bootleggers, who sell alcoholic beverages illegally; they get the market to themselves on Sundays. Baptists and other religious groups support the same laws, but for entirely different reasons. They are opposed to selling alcohol at all, but especially on Sunday. They take the moral high ground, while the bootleggers persuade politicians quietly, behind closed doors.

Such a coalition makes it easier for politicians to favour both groups. The Baptists lower the costs of favour-seeking for the bootleggers, because politicians can pose as being motivated purely by the public interest even while promoting businesses' interests.

The post-Kyoto period promises to be rich with bootlegger–Baptist coalitions. The Baptists are the environmental groups pushing for ratification and enforcement of the treaty, and working to prevent backsliding.[38] They are passionate and persuasive to the public as they argue that cutting back on carbon emissions is a moral necessity. Robert Nelson has noted that the environmentalist movement shares many characteristics with Calvinist Puritanism, namely, a view of mankind as 'deeply sinful', a view of the world as corrupted by greed and sin, and a belief that the remedy is to renounce man's sinful ways for a more 'pure' lifestyle.[39] Philip Stott has explained: 'In Europe, "global warming" has become a necessary myth, a new fundamentalist religion, with the Kyoto Protocol as its articles of faith.'[40]

To determine which groups are the bootleggers, we should search for special-interest groups who are positioned to gain from regulatory enforcement and stringency or who must fend off losses that spring from proposed rules. Some countries, such as the UK, are positioned to exploit carbon reductions they have made in the past by raising the cost to economies that still rely heavily on coal. European nations can effectively rely on the 'bubble' proposal to keep their own compliance costs lower relative to other countries not favoured with such a system. Developing countries, with their own emissions unlimited, see opportunities for payments from industrialised countries for reducing carbon emissions or for planting trees. Moreover, within countries, some industries are favoured by the rules and, within industries, some firms will be favoured. Environmental activists provide the cover story on which media attention is focused, while companies, industries and countries work quietly in the background to gain benefits.

The battle over tradable permits: the implications of bootleggers and Baptists

In the dismal science of economics, perhaps nothing is so universally accepted as the superiority of price incentives and property rights measures, as opposed to regulation, to achieve a goal such as cutting emissions.

Regardless of the difficulties that might arise in implementing a global tax or global permit trading system, both are theoretically superior to command-and-control standards. But an important implication of the bootleggers-and-Baptists theory is that global environmental regulation will overwhelmingly prefer flat standards, rather than taxes or tradable permits.

Despite (or because of) the fact that direct taxes or tradable permits are more efficient, industries usually prefer standards. As Stephen

Breyer notes, standards have two primary anti-competitive effects – raising barriers to entry and imposing disadvantages on smaller firms (who are less able to attain economies of scale).[41] Because the firms to be taxed or regulated are a more concentrated special interest than the public (who would benefit from the tax proceeds), the firms naturally are able to influence political decisions.

Environmental activists tend to prefer standards over taxes or tradable permits as well. This is because of the moralistic attitude they adopt, in keeping with the Baptist role. Baptists, after all, would not be content with a mere tax on Sunday sales of liquor. *Any* liquor sales on Sunday are deemed immoral and should be blocked completely. Similarly, a prohibition on emissions over a certain level appeals to the environmentalist moralist more than a tax, which would allow any given corporation to emit as much as it pleased if it pays the tax, and indeed, would allow a corporation to increase emissions if it finds a cost-saving technology.[42]

Similarly, in the international context, poorer nations tend to see tradable permits as inherently unfair.[43] If an international permit trading scheme is enacted, it will probably also face the allegation that rich countries are using it to exploit poor countries.[44]

Thus, the bootleggers and Baptists theory explains why regulatory standards are often the result of political bargaining, as opposed to the more efficient solution of penalty taxes. Both firms and environmental activists effectively collude in lobbying for the most inefficient form of regulation.

Even though unfettered permit trading would reduce the overall cost of controlling carbon emissions (assuming that a viable international market could be constructed), the prospect of having the USA reduce its costs was apparently more than some European politicians could bear. In earlier rounds of negotiation, British Deputy Prime Minister John Prescott expressed concern that Washington would 'buy tradable greenhouse emission permits from Russia'. As he put it, 'Europe has always been clear that while we accept the trading possibilities in this matter, they should not be used as a reason for avoiding taking action in your own country.'[45]

One economist has suggested that Europe's high energy taxes have stifled economic growth, and that Europe wants to 'force countries like the United States who have relatively low energy taxes to give up their competitive advantage in this area'.[46]

Ironically, the EU's opposition to unlimited permit trading would result in harm to smaller and poorer nations that could profit from selling permits to larger, wealthier countries. According to one analysis, five developing countries (Thailand, Pakistan, the Philippines, Korea and Vietnam) stood to gain US$6.1 billion a year from selling certified emissions reductions to rich countries, versus US$1.4 billion if tradable emissions reductions were limited to 25% of the rich countries' reductions.[47] Yet the fact that the producers' surplus would also be reduced by unfettered trading[48] was apparently enough to make the EU want to impose supplementarity requirements.

Long before Kyoto, academic economists churned out vast numbers of studies on the effects of controlling GHGs. During the negotiation of Kyoto, these broad studies were supplemented by studies of its impact on specific sectors. There are three key observations that can be made about numerous studies done so far. Most studies predict that the cost to the USA of ratifying the Kyoto Protocol would have been quite significant. They also indicated that the Kyoto Protocol would have had widely disparate impacts on various industries, but that it would have quite different effects on different countries. We will examine these findings in turn.

More pertinent to the thesis of this chapter are the findings about how Kyoto would impact different energy industries. Jorgensen's and Wilcoxen's simulations indicated that meeting Kyoto's goals would require that coal be taxed at US$11.01 per ton, oil at US$2.31 per barrel, and natural gas at US$0.28 per thousand cubic feet.[49] As a result, they estimated that coal prices would increase by 40% and coal production would drop by 26%.[50] Manne and Richels found that under the most stringent Kyoto assumptions, the price of coal would increase fourfold, and the demand for oil would increase, not decrease.[51]

DRI–McGraw-Hill's study (reviewed by the Economic Policy Institute) assumes a government-issued marketable permit programme.[52] The study considers two scenarios: stabilising GHG emissions at 1990 levels and reducing emissions by 10% below 1990 levels by the year 2010.[53] The study predicts that various energy forms will see severe price increases: a sevenfold price increase for coal by 2010,[54] a doubling of electricity prices and retail gasoline price increases of 40–50%.[55]

Because of these price increases, coal, which now provides 24% of

US energy, would provide only 18%.[56] Petroleum's share would increase, and natural gas would maintain its current market share.[57] Coal output was predicted to decline by 45%, rubber and plastics by 50% and electricity production by 18%.[58]

WEFA's study found that to achieve the necessary reductions, carbon permit prices would have to rise by an estimated US$265 per metric ton, causing an increase of 65 cents in the cost of a gallon of petrol and a doubling of natural gas and electricity prices.[59] Industries producing chemicals, paper, textiles and apparel, and computer and electronic parts production would be severely affected.[60]

Kyoto would also affect countries in quite different ways. First, countries will have different marginal costs of abatement, which means that their costs to comply with Kyoto will vary tremendously. Unfortunately, few studies examine marginal costs of abatement on a country-by-country basis. An educated guess would be that countries that have already substantially reduced greenhouse emissions, which have already passed environmental regulation internally, and which have already invested in energy efficient products, would probably have higher marginal costs than those countries where energy emissions have heretofore been relatively unregulated.

Second, industries are not evenly distributed among countries. Given the differential impact that Kyoto will have on various industries, countries would be affected to the extent they possess more or fewer of those industries. Any decrease in oil use, for example, would be opposed by countries that depend heavily on oil sales for their national income.

Third, the exemption for developing countries could have a distributional effect on where industries locate. Manne and Richels note that under Kyoto, 'US output of energy-intensive products such as steel, paper, and chemicals could be 15% less than under the reference case by 2020 ... In contrast, countries such as China, India, and Mexico would increase their output of energy-intensive products.'[61] WEFA found that, due to the exemption for developing countries, US exports would become 'relatively more expensive on the world market', while the prices of many imported products would fall.[62]

Corporations as bootleggers

The arena of environmental regulation is rich with opportunities for favour-seeking, bootlegging and exploitation by self-interested corporations on all sides of the global warming debate. Industries or corporations that expect to benefit from the regulatory standards imposed by Kyoto lobbied in support of the agreement, while those that would be harmed opposed it.

The squabble over subsidies

One of the most potent arenas for rent-seeking involves government subsidies to industry for research and development on global warming projects. Government subsidies are usually inefficient because subsidies should go only to projects that 'promise great potential gains for society but are unlikely to yield profits to the innovator'.[63] Yet government is often unwilling to restrict its funding to such projects, if it is able to identify them in the first place. As a result, government R&D subsidies probably have the effect of crowding out private efforts that would already occur.[64]

Ironically, government subsidies have created opportunities for rent-seeking on both sides of the global warming controversy. Though it may not be well known, the US government already subsidises the very oil companies that are seen as greenhouse villains by environmental activists.[65] One study claimed to have found that between 1992 and 1998, two US agencies – the Overseas Private Investment Corporation and the Export-Import Bank of the United States – underwrote some US$23 billion in financing for oil, gas and coal projects throughout the world.[66] The study claimed that these projects will, over their lifetimes, release 29.3 billion tons of CO_2, more than all global emissions in 1996.[67]

But the global warming scenario has also created opportunities for corporate welfare. President Clinton's budget for the fiscal year 2001, for example, asked for US$1.4 billion for research on 'clean technologies for the buildings, transportation and electricity sectors', US$200 million to 'make the latest energy technologies available to the developing world', and US$1.7 billion for research on the role of GHGs in global warming.[68]

One thing is sure, however – tax breaks, subsidies and regulations requiring energy-efficient products are sure to be supported by the manufacturers of those products. For example, Exxon-Mobil, a traditional oil corporation, has published opinion advertorials urging the government to fund research on such technologies as fuel cells, which happens to be one of Exxon-Mobil's corporate projects.[69] This is why prominent economists argue that a simple carbon tax would be preferable to subsidies, because subsidies 'give too much power to government officials who pick the favoured activities'.[70]

Enemies of coal and oil

Supporters of Kyoto include not only those companies which are directly subsidised, but those which offer products that compete with coal or oil, and that would therefore benefit from restrictions on carbon emissions. As reported in *US News & World Report*, 'Many businesses active on global warming envision not subsidies but a market-based trading system that would allow farmers and others who cut carbon emissions to get credits they could sell to carbon-emitting businesses.'[71]

The nuclear power industry, for example, strongly supports Kyoto, just as it supports any policy that will make nuclear alternatives more expensive.[72] Nuclear plants emit no CO_2, and London's Uranium Institute has said that nuclear plants currently cut carbon emissions by 2.3 billion tons per year worldwide, despite supplying a mere 6% of the world's power.[73]

But nuclear energy is in trouble throughout much of Europe, as many countries (including Belgium, Germany, the Netherlands and Sweden) have decided to phase out nuclear power.[74] Because Kyoto would pressure those countries to limit carbon emissions, it would compel them to continue using nuclear power.

Indeed, without nuclear energy, many countries would probably be unable to meet their Kyoto commitments. EU Energy Commissioner Loyola de Placio has said that without nuclear energy, 'we won't be able to stick by the terms of the Kyoto agreement.'[75] Similarly, Germany's Economy Minister Werner Mueller has said that because of his country's phase-out of nuclear power, 'a CO_2 reduction of 40% by 2020 is hardly possible.'[76]

All of these considerations mean good times ahead for the nuclear industry, which 'sees global warming as its trump card'.[77] As the International Energy Agency said in a recent report, 'A strong commitment to reduce emissions of CO_2 could have a dramatic positive effect on the prospects for nuclear power over the coming decades.'[78]

The nuclear industry's claims to environmental friendliness are challenged, of course, by environmental activists, who have lobbied to keep nuclear energy from being an option for reaching the Kyoto reductions.[79]

Not everyone is teaming up with the Baptists. Many major industries, or at least major firms, still oppose Kyoto. Coal producers and related unions have been among the most vocal in their opposition. Coal interests in West Virginia, for example, successfully obtained state legislation which prohibits the state government from 'proposing or enacting rules regulating so-called GHG emissions from industrial sites'. Yet even this anti-Kyoto legislation was announced with a flourish of Baptist-like rhetoric. When signing the bill, Governor Cecil Underwood said that while actions like the Kyoto Protocol must be opposed, we 'should continue to encourage the development and implementation of technologies that allow the clean burning of coal'.[80]

Countries as bootleggers and Baptists

Like any global environmental treaty, Kyoto's negotiations offered many opportunities for countries to compete with each other. In the words of *The Economist*, the treaty 'always had more to do with the jockeying for individual trading advantage than preserving the global environment for future generations'.[81]

The first thing that appears to be a bootlegging measure is the choice of 1990 as the baseline year for measuring any emissions reductions.[82]

As reported in the *Oil and Gas Journal*:

> Against 1990 baseline levels, European Union members seemed to have shouldered the greater load, agreeing to an aggregate 8% reduction in greenhouse-gas emissions against 7% for the US. But several European countries displaced coal significantly with natural gas between 1990 and the 1997 meeting in Kyoto. With this convenient head-start, the EU as a whole needed to lower emissions from levels projected for 2012 by only 15–20% to satisfy Kyoto, compared with 30–35% for the US.[83]

On the other hand, countries such as Canada, the Netherlands, Australia and the USA had increased carbon emissions between 1990 and 1997 – which meant that their burden under Kyoto would be comparatively much larger.[84] Thus, even in something as seemingly innocuous and objective as the choice of a baseline year, various countries attempted to set Kyoto's terms so as to maximise their competitive advantage.[85]

Redistribution effects

In the international context, the 'Baptists' are the tropical countries that have urged the ratification of Kyoto on the grounds that their lands would be flooded and their agriculture destroyed if global warming occurs. As reported in the *New York Times*, a study by the Intergovernmental Panel on Climate Change argued that global warming would further widen 'the gap between rich, industrialised countries and poor developing nations'.[86]

While poorer nations proposed the moralistic, Baptist arguments, probably a more important motivation was their desire to create incentives for job and wealth redistribution from wealthier countries to poorer countries. Due to the exclusion of developing countries from Kyoto's restrictions, one predicted effect is that industrial production and jobs might shift from the industrialised world (where energy prices would be higher) to the developing world.[87]

This is because the agreement would raise the relative cost of production in the developed world, particularly for manufacturing and energy-intensive industries, encouraging a reallocation of investment to less developed countries.[88] Thus, though Kyoto was promoted as an environmental treaty, it might well have functioned as a redistribution policy from rich countries to poor countries, even as those very same poor countries played to Baptist moral sensibilities by presenting themselves as most vulnerable to the impacts of global warming.

Some environmental leaders openly acknowledged this aspect of Kyoto; Margot Wallstrom said in reaction to Bush's rejection of Kyoto:

> This is not a simple environmental issue where you can say it is an issue where the scientists are not unanimous. This is about international relations; this is about trying to create a level playing field for big businesses throughout the world. You have to understand what is at stake and that is why it is serious.[89]

In this praiseworthy moment of honesty, the European environment commissioner laid bare the real motivations behind the EU's outrage over the USA's actions: 'international relations' and creating a 'level playing field for big businesses throughout the world' – in other words, bootlegging on a global scale. The merits of such a policy are at best debatable, and should be discussed openly rather than hidden behind moralistic rhetoric.

The battle over sinks and tradable permits

In the November 2000 negotiations at COP-6, the United States and the EU battled over how to count reforestation and other methods of creating carbon sinks.[90] Specifically, they disagreed over how much trading in emissions rights and how much credit for forestation would be allowed.[91] The USA, for example, wanted to get 'sink' credit for a mere 20% of the estimated 288 million tons of carbon absorbed each year by US forests.[92] This is not surprising – one recent article in *Science* estimated that all the carbon emitted by the USA and Canada was balanced by absorption in forests and vegetation.[93]

But the EU tried hard to limit any credit for carbon sinks,[94] at least in part out of jealousy because their own countries have less land available for reforestation efforts.[95]

The EU was also inspired by a desire to punish the USA for not having enough market-stifling command-and-control regulation. *The Economist* reported, 'Some European ministers made it clear that they wanted Americans to feel some economic pain more than they wanted a workable agreement.'[96] The Clinton administration's chief Kyoto negotiator agreed, saying

> The EU is concerned that implementing the Protocol, particularly in the United States, will be too easy. Some in Europe think that we have a moral obligation to change our lifestyle as quickly and radically as possible. In this sense, many in the EU believe that producing significant short-term pain and suffering is actually desirable, rather than something to be avoided. The EU is also concerned that enterprises in the United States and other countries relying on efficient market-oriented approaches will enjoy a competitive advantage over European businesses that have been subjected to carbon taxes and extensive regulation.[97]

Another battle was about the extent of permit trading to be allowed. Several countries, including the former Soviet-bloc countries and Russia, stand to profit hugely from a permit trading system if Kyoto is implemented, so it is no surprise that they strongly support Kyoto. The Soviet-bloc countries experienced an economic and industrial collapse in the 1990s, which reduced their carbon emissions by some 40% from their 1990 levels. They need not take any action to reduce their existing emissions, and because the Kyoto Protocol would increase the price of carbon, they could profit handsomely by selling carbon permits that they would not need to use domestically.[98]

Meanwhile, Russia's economy is in shambles – production has fallen 40% below 1990 levels, which means that they are not really selling emissions 'reductions' at all.[99] Selling these unused emissions credits could be a huge boon: Kazakhstan could earn some US$800 million per year, and Russia some US$3 billion per year, if Kyoto is ratified, assuming current carbon prices.[100] If, as most analysts expect, the price for carbon under Kyoto rose to around US$20 per ton, Kazakhstan could earn US$3 billion and Russia US$12 billion per year.[101]

Other countries were not so keen to enter the world of emissions trading. The EU in particular was suspicious of allowing too much permit trading, instead pushing for supplementarity requirements that could only reduce the available gains of trade for all parties.

But even as the EU tried to limit gains from trade for the USA, it created a way for its member states to minimise their own emission reduction costs. The EU would employ a 'bubble' concept to achieve overall emission reductions for the member states.[102] Under this plan, some European countries will be able to emit more and others less because only the collective total matters.

The bubble allows the EU to minimise the overall cost of emission reductions by allocating emission cutbacks differently to different countries. It is generally cheaper to reduce emissions when concentrations are higher, so it is logical and technically efficient to require countries that produce more CO_2 per unit of output to make larger cutbacks. The bubble system would probably save vast sums of money for the EU.

The system obviously provides a framework for trading. Those which face high costs in cutting emissions can buy permits or credits from other European countries that face lower control costs. By encouraging bubble trading within the EU while managing Europe's external trades so that its competitors' costs go up, Europe's new central government takes on the traditional protectionist position of many nation-states, controlling exports and imports. The difference is that the items traded are permits (emission reductions), not commodities.

Countries within Europe are strategically positioning themselves

188 Adapt or Die

against each other. The bubble policy gives more leeway to southern European countries, such as Spain, Portugal and Greece, lower-income countries that are rapidly industrialising and emitting high quantities of carbon.[103] Countries such as Germany, the UK, the Netherlands and France already had relatively low carbon emissions, having transformed their coal-based energy economies to cleaner fuels. Through trading, these countries can purchase the emissions credits of the high carbon emitters at relatively low cost.

Not surprisingly, the USA's negotiators opposed the European bubble when it was proposed during the COP-6 negotiations.[104] But in its own way, the USA was playing the 'increase the rivals' cost game'. American negotiators complained that the EU 'bubble doesn't level the playing field',[105] and argued that each European state should have specific reduction goals that must be met internally, rather than being allowed to trade within Europe under the bubble.[106] From the USA's perspective, this arrangement would have been doubly beneficial: the USA could trade with less-developed European countries, and at the same time impose higher costs on its competitors.

Our analysis of the bubble scheme suggests newer EU members received more allowances, all else being equal,[107] and that the allocations are designed to keep the bootlegger community intact. The bubble minimises overall costs and other concessions (sometimes called 'side payments') are made to ensure that reluctant community members do not trade outside the community. Populous nations with tighter emissions allowances will probably buy permits from the countries with higher carbon streams and larger allowances for emission growth. Wealth will flow generally from northern to southern European countries for trades within the European bubble.

Countries and industries
One obvious effect of Kyoto is to provide a disincentive for oil use. It is thus unsurprising that some of Kyoto's most vocal opponents were those countries that depend heavily on oil sales.

Saudi Arabia's energy minister, for example, said, 'We cannot accept that the industrialised countries transfer the cost of reducing their GHG emissions to our countries by embracing policies and measures that would lead to reducing their imports of our fossil fuel exports on which revenues we depend to a great extent.'[108] He also said that if Kyoto were ratified, the OPEC countries would lose US$60 billion of income annually by 2010, more than one-third of their current revenue.[109]

Enforcement problems: further possibilities for bootlegging

A serious problem is that Kyoto would be very difficult to enforce. Even if Kyoto were enacted, bootlegging would have continued in the attempts to enforce the treaty. While some reports indicate that the EU favours a system of fines for excess carbon emissions,[110] it would be extraordinarily difficult to enforce such fines as long as individual countries are (as they must be) responsible for measuring and estimating their own emissions.[111] Moreover, GHG emissions are an estimate rather than a precise measurement.[112]

Thus, Kyoto's limits would 'require very complicated calculations, and government agencies, industries, and environmental groups will expend large amounts of time, talent, and political capital trying to influence how experts calculate those estimates'.[113] Perhaps an international environmental police force would be required, although the political likelihood of that happening is virtually non-existent.

Even if Kyoto's terms could be enforced with ease, another problem would remain: 'only about half of greenhouse gas emissions have come from burning fossil fuels.'[114] The other half of human emissions are impossible to monitor, because they result from pipeline leakage, the burning of tropical forests, wood-burning for fuel, and other sources that are impossible to monitor.[115] And as they are more likely to occur in developing countries, omitting them from Kyoto coverage would 'probably favour' those countries.[116] Thus, difficulties in enforcement create further opportunities for bootlegging on an international scale.

Conclusion

A world industrial policy is in the making. In the past, socialist and communist governments (and even, to a lesser extent, the USA) engaged in industrial planning within their countries. The publicly stated goal was to improve economic well-being by favouring certain industries and firms to be the engines of the economy and allow the others to phase out gradually. Such centralised planning was never effective in the long run, but it created opportunities for favour-seeking that gave some industries and firms advantages over the others.

Day after day, newspapers and television continue to report the alarmist pleas of the 'Baptists', urging world leaders to 'do something' about global warming, while the machinations of the 'bootleggers' largely go unnoticed. Yet there is ample evidence that the Kyoto Protocol (or any similar agreement) would be used as a crutch to secure political favours for conventional special-interest groups. As we have seen, some nations and at least one community of nations dictated Kyoto's terms in strategic ways to enhance their positions relative to other nations.

In the final analysis, we should hope that the fear of perilous global warming will subside, along with efforts to control the world's energy economies. Yet even if this happens, the regulatory concrete delivered by Kyoto would endure. History shows that once a major concern becomes transformed into institutional rules and regulations, interest groups that invested in those rules seek to maintain them.

The Kyoto Protocol might have established a system of industrial policy as well, although its purpose would not be to achieve economic growth. The officials in charge of the system would have possessed the power to specify which nations and industries would bear the greater pain of cutting back on carbon emissions. In this international system, although the Baptists presented the moral front for adopting the treaty, the bootleggers would have converted environmental policy to an industrial policy that favours them.

If we discover that global warming isn't such a problem, the treaty will be a waste of time and a misuse of our resources. If global warming turns out to be genuine, those economies that maintain market flexibility will be best equipped to adapt to it.

This chapter originally appeared as a longer article in the *Harvard Environmental Law Journal,* vol. 26, no. 1, 2002, pp. 177–229. This edited version is reprinted here by permission.

10 Climate change and civilisation collapse

Dr Benny Peiser

> Like most things, collapse explanations are subject to fashion, and
> the one most in the limelight today is climatic change ... Right
> now mega-drought is the 'hot' explanation for the Classic [Maya]
> collapse, and the usual bandwagon effect is in full career among
> many of my colleagues, although others remain properly
> suspicious of drought as the triggering mechanism.
>
> David Webster, *The Fall of the Ancient Maya*, London 2002

Introduction

One of the most powerful drivers of environmental gloominess and
cultural pessimism is the spectre of ecological apocalypse. The muta-
tion of age-old, religious end-time prophecies into secular predictions
of natural cataclysms and societal collapse – in short, the emergence
of environmental apocalypticism – is perhaps the most significant ideological development in the western world since the demise of Marx-
ism.

Marxist doctrine, let us never forget, crumbled because its predicted, and eagerly anticipated, disintegration of free market
economies never transpired, but communist economies and totalitar-
ian dictatorships have mostly come to sticky ends. Deeply infuriated
by the failure of their predictions and the unremitting vibrancy of cap-
italism, many disillusioned believers turned to ecological pessimism
and environmental determinism. Not for the first time in the long history of apocalyptic movements, new wine was poured into old bottles.

Many ideologues replaced their old beliefs in economic decline and breakdown with the new principle of ecological decay and disaster.

There is no shortage of physical factors that can produce natural disasters and social deterioration. These could include catastrophes due to asteroid and comet impact, the failure of global agriculture due to volcanic super-eruptions, the reappearance of a new ice age, epidemic diseases, etc. However, none of these horror scenarios has alarmed the public as much as the alleged peril of human-caused global warming.

Originally, this idea was a theoretical speculation about the supposedly negative impact of increasing CO_2 emissions into the earth's atmosphere. Recently, these speculations have turned into a veritable scare. We are warned that if we fail to drastically reduce CO_2 emissions, this will cause global warming which will trigger social upheaval and natural disaster everywhere. Some of the experts on civilisation collapse have argued that dealing with climate change 'will require substantial international cooperation, without which the 21st century will likely witness unprecedented social disruptions'.[1]

This chapter argues that environmental determinism is the latest fad in explaining past societal evolution and civilisation collapse. It is beyond doubt that a number of complex societies fell apart during the last 4,000 years. Climate change is one possible explanation for those collapses, but no one has identified the basic dynamics or driving forces of societal dissolution. Warmer periods have had a considerably benign role in social, economic and technological progress, but global cooling and cold spells have been largely detrimental to societies. Today, in contrast, we have the technological capacity to deal with climatic changes, even with events that may have been catastrophic in the past.

Historical evidence for civilisation collapse

To further amplify the alarm that a warming climate would lead to collapse of human civilisation, a number of researchers have turned their attention to historical examples of societal breakdown and have incriminated 'climate change' for the disintegration of ancient societies.[2] From the downfall of the Akkadian civilisation to the demise of the ancient Mayans and the Roman Empire, climate change is increasingly named as the cause of a number of ancient civilisation collapses.[3]

The discovery of warming and cooling cycles during the Holocene – the present geological era – has drawn attention to historical climate events such as the Little Ice Age, a prolonged cold spell that affected wide areas of the globe for several hundred years.

There is increasing evidence that the climate in the Holocene era has been much more variable than once believed. Abrupt climatic downturns in the form of temperature decreases appear to have been significant enough to cause agricultural disruptions and other adverse effects in a number of historical cases. Warmer temperatures, on the other hand, have never contributed to the decline or disintegration of any society. Climate alarmists consistently fail to mention that warm periods during the Holocene have played a considerably benign role in social, economic and technological progress. Global cooling and cold spells, on the other hand, have been largely detrimental to society's advance and evolution.

Throughout much of human history, cities, regions and entire states have come and gone. The stability of agriculture-based societies has always relied on the constancy of climatic conditions. Archaeological research has unearthed heaps of empirical evidence for the inherent vulnerability of agricultural societies to climatic change. Since agricultural production relies on factors such as fertile soils, precipitation or irrigation, prolonged climatic downturns, droughts or desiccation have regularly led to mass migration and resettlement. Some regions which were once inhabited 5,000 years ago (such as parts of the Sahara) today are deserts – barren and abandoned.

But relocation of a society does not necessarily mean the unmitigated 'collapse' of a society. Neither are episodes of climate changes detrimental to all, or negative in all of their aspects. If a social group abandons terrain which has become sterile, it is does not mean 'societal failure' if they move to more fertile territory. In fortunate circumstances, the movement of social groups to new territory is beneficial and can lead to higher levels of societal complexity.

The desertification of the Saharan region 7000 to 5000 years ago, for example, set in motion a significant advance in societal evolution. It is now generally thought that the environmental pressure behind these events may also have been the main impulse that led to the foundation of complex urban civilisations along the Nile, Euphrat and Tigris rivers.[4]

In short, the evolution of society during the Holocene has been marked by recurrent patterns of expansions and downturns. Some cultural declines have been gradual, occurring over centuries, and others have been more abrupt. Warfare, power struggles, diseases, over-population, economic disruptions, droughts, or natural disasters can facilitate a breakdown in social order and a decline in cultural complexity. Internal causes (such as political conflicts or over-farming) can combine with external causes (such as war or natural disaster) to bring about malfunction and failure.

There is unambiguous evidence that a number of urban civilisations which had once acquired high levels of social complexity (such as the Akkadian, Roman and Mayan civilisations) declined and disintegrated. Yet despite burgeoning research and mounting data, no one has identified the basic dynamics or driving forces of societal dissolution. What we know for certain, however, is that every age has its own favourite explanations for the collapse of civilisations before it. The latest fad for such explanations is environmental determinism.

What wrecked the Roman Empire in the West?

Consider the decline and fall of the western Roman Empire. For centuries, scholars have mulled over the possible reasons for its demise. According to traditional assumptions, barbarian invasions which ransacked Rome in AD 476 brought about the Empire's fall. The end of classical civilisation came 100 years later after an epidemic of bubonic plague swept the region, and invasions by Slavic tribes brought wide-scale destruction.

Ancient authors attributed the social decadences and Roman decline to mounting bureaucracy and excessive taxation. Writers during the Enlightenment (Edward Gibbon, for instance) favoured moral and religious explanations and blamed Christian anti-paganism for Rome's downfall. Marxist historians, on the other hand, preferred economic and social factors, blaming class conflicts, political struggles and imperial over-stretch.[5] The emergence of the environmental movement illustrates that the focus has shifted yet again. Rather than considering internal and social factors, environmentalists prefer ecological explanations, blaming population growth, environmental degradation, deforestation or 'climate change'.[6]

Around AD 540, parts of Europe did indeed experience rapid cooling. This period corresponds with worldwide accounts of a significant climatic downturn due to a mysterious dust-veil event.[7] The cause of this cooling is still unknown, but some researchers speculate that it was either the result of a massive volcanic eruption or due to some cosmic dust loading of the stratosphere. Tree-ring data from Europe and North America indicate a significant temperature drop around AD 536. They also show that the tree-ring widths returned to pre-AD 535 scale in the late 540s, suggesting that the climatic downturn lasted for some fifteen years.[8] Other research suggests that the cold period began as early as AD 500 and lasted for more than 200 years.[9]

Evidently, there is no consensus about the duration or cause of the 'European Dark Ages Cold'. Nor is there any agreement about whether the Roman Empire tumbled because of a climatic downturn, due to political and economic discord, or as a consequence of a multitude of factors. After all, the onset of a cool period did not lead to the crash of the Byzantine Empire, which survived for another 500 years. Thus, we remain in the dark about the real reasons why Europe's classical civilisation ended in the West but continued in the East.

The fall of the Akkadian Empire

The Akkadian Empire dominated large parts of Mesopotamia during the late third millennium BC. At some time between 2350 BC and 2200 BC, it disintegrated. Traditional explanations for its demise range from warfare and internal rebellion to socio-political dissolution. In recent years, 'climate change' has become the most popular theory not just for the cessation of the Akkadian Empire, but for the seemingly simultaneous collapse of other urban civilisations around the world.

A number of researchers are convinced that in 2200 BC, an abrupt drop in global temperatures occurred that disrupted the weather around the globe.[10] One of these researchers, Yale University archaeologist Harvey Weiss, claims:

> The Akkadian empire of Mesopotamia, the pyramid-constructing
> Old Kingdom civilisation of Egypt, the Harappan 3B civilisation
> of the Indus valley, and the Early Bronze III civilizations of
> Palestine, Greece, and Crete all reached their economic peak at
> about 2300 BC. This period was abruptly terminated before 2200

BC by catastrophic drought and cooling that generated regional abandonment, collapse, and habitat-tracking.[11]

Abrupt climatic downturns are not unusual since they repeatedly occur in response to significant volcanic eruptions and other atmospheric dust-loading events. The abrupt cooling in 2200 BC, however, is thought to have disturbed agricultural production around the globe for three centuries. This cooling is assumed to have triggered catastrophic sandstorms and mega-droughts, subsequently prompting societal disintegration on a global scale.[12]

But how compelling is the case for 300-year-long sandstorms, droughts and climatic disaster? What physical mechanism could have triggered such a global calamity? So far, there have been no convincing answers to these questions. In fact, there remain grave doubts within the scientific community about this entire theory, particularly with regard to its magnitude, its physical causation and its chronology.

For a start, archaeological evidence seems to be irreconcilable with its basic premises. Fifty years ago, French archaeologist Claude Schaeffer discovered that most Near Eastern settlements had been repeatedly destroyed and abandoned throughout the 2000 years of Bronze Age cultures (~3000–1200 BC). In short, the recurring destruction and abandonment of Bronze Age settlements is the norm, not the exception, throughout the Near East. There also seems to be archaeological evidence for seismic activity at the end of the Early Bronze Age, which would be incompatible with a simplistic climate model of civilisation collapse.[13]

Even more doubts have been raised about the chronology of the events in question. Marie-Agnès Courty, a leading team member of the Tell Leilan excavations[14] which formed the basis of the original idea, has recently revised the dating of the 'abrupt' onset of environmental change to 2350 BC, thus shifting the date by some 150 years.[15] Butzer[16] and Baillie,[17] two of the world's leading paleo-environmental researchers, have highlighted the inherent imprecision of the dating methods used.[18]

In any case, the claim for an abrupt worldwide climatic downturn in 2200 BC is difficult to sustain if such a global disaster does not show up in the most sensitive and reliable climate indicator, Californian and European tree rings! Critics have thus underscored the elusiveness of any conclusions drawn from a still unreliable chronology:

Are we looking at an event starting in 2400 or 2350 or 2200 or 2180? Butzer with his '2400–1900 BCE' has broadened the debate to a full half millennium – a time period so long that we could reasonably expect some environmental changes to be recorded in most areas.[19]

Other researchers have pointed out that certain civilisations, such as the Bronze Age culture of the Mediterranean island of Crete, were not at all affected by the hypothetical climate disaster.[20] In fact, a number of cities in the Near East and on the Indian subcontinent appear to have expanded and progressed to higher social complexity during the nadir of the supposed mega-drought.[21]

Furthermore, advocates of the climate catastrophe theory have not yet presented a coherent physical model that would provide an explanation for an abrupt climate crash which persisted for 300 years. Even American tree-ring data, which is cited as confirmation of the global scale of the prolonged catastrophe, only shows a slow and gradual decrease in temperatures which returned to the pre-cold average in 2056 BC.[22]

Regardless, many Akkadian settlements were abandoned after 2200 BC while people moved south, where irrigation-fed agriculture continued. While the Akkadian state and its imperial bureaucracy crumbled, elements of the culture itself relocated and lived on. And not everyone fled the area since nearby settlements continued their existence at diminished levels. Nevertheless, the evidence for some sort of natural calamity seems undeniable. While no one knows what brought it about, its overall societal effects on Mesopotamia and the Near East – let alone the rest of the world – are far from certain.

The Maya civilisation is dead, long live the Maya civilisation

The decline and fall of the Classic Maya civilisation around AD 800 is another popular case in point – climate change is blamed for its demise. Modern-day ecological concerns such as population growth and environmental degradation are frequently cited as the cause of the Maya 'collapse', but climate change has certainly become the most popular culprit in recent years.

Recurrent droughts and mega-droughts on the Yucatan Peninsula of Mexico have been a persistent feature of Mesoamerican civilisations for centuries. There is a wealth of paleo-environmental evidence that local and regional climates have varied considerably during the last 10,000 years.[23] In addition, there is ample evidence that Maya migrations, the abandonment of settlements and societal 'mini-collapses' were recurring features of a highly volatile culture.

According to a number of researchers, the event which triggered the Classic Maya 'collapse' was an abrupt climatic incident and a prolonged drought that began early in the ninth century AD. In keeping with this scenario, food production dropped drastically as a result of the mega-drought, resulting in large-scale famines. Ensuing endemic diseases caused plagues and epidemics. In view of such chaos, the disintegration of social and political structures most probably contributed to the breakdown.[24]

The climate change hypothesis of Maya collapse includes a variety of supposed causes, factors and effects. This multitude of suggestions is not entirely new. It is actually characteristic of research on civilisation collapse. As early as the 1930s, two pioneering Mayanists published a list of hypothetical explanations for the demise of the Classic Maya which included climatic change, exhaustion of the soil, epidemic diseases, earthquakes, war, national decadence and religious superstition. As David Webster points out, even after 70 years of painstaking research, these are still the same basic assumptions for its termination.[25] Indeed, one Mayanist has claimed to have counted as many as 100 distinct theories, explanations and hypotheses for the Maya 'collapse'.[26]

More recently, the mega-drought hypothesis has received further support with research that provides additional evidence of a hundred-year drought punctuated by shorter but multi-year droughts at around 810, 860, and AD 910.[27] According to this theory,

> rapid population expansion during a climatically favourable
> period from about 550 to 750 AD left the civilisation operating at
> the limits of the environmental carrying capacity. This left the
> Mayan society highly vulnerable to subsequent multi-year
> droughts and led to its collapse.[28]

The very term 'civilisation collapse' may be altogether inappropriate, however, or at least misleading. The Maya culture was neither finished for good, nor were all of its settlements abandoned at the end of the

Classical period. Some regions show clear evidence of the continuity of Maya occupation and culture. These areas were mainly located in the northern plain of Yucatan and inland around a chain of lakes. Many Maya migrants, it would appear, fled the droughts and resettled where water sources were still readily available. It was in these settlements that the Spanish discovered a highly complex Maya society when they arrived in the sixteenth century.

The fall of complex and the rise of hyper-complex civilisations

Whatever the details of societal decline and disintegration in antiquity, it is beyond doubt that a number of complex societies fell apart during the last 4000 years. Such temporary breakdowns, however, have never been unmitigated or total. Most unsuccessful ancient societies recovered after a period of marked decline and regularly emerged more robust and dynamic. After all, the general trend of cultural evolution during the last 10,000 years has not been intermittent breakdown of societies but relentless technological progress, increased social complexity and much improved safety measures against the forces of nature.

The imagery of powerful civilisations breaking down has nonetheless contributed to a mindset which is increasingly fretful about the fate of our own civilisation. All too often, these worries are based on misleading analogies with agricultural societies that were especially vulnerable to environmental stress and lacked the benefits of modern technologies to cope with changes.

According to climate alarmists, predicted global warming will cause even more unstable conditions in our modern world which might be populated by up to 10 billion people in a few decades. This reasoning suggests that today's 'overcrowded earth' may hamper such adaptation:

> The magnitude of expected temperature changes gives a sense of
> the prospective disruption. These changes will affect a world
> population expected to increase from about 6 billion people today
> to about 9 to 10 billion by 2050. In spite of technological changes,
> most of the world's people will continue to be subsistence or
> small-scale market agriculturalists, who are similarly vulnerable to
> climatic fluctuations as the late prehistoric/early historic societies.

Furthermore, in an increasingly crowded world, habitat-tracking
as an adaptive response will not be an option.[29]

Contrary to popular belief, the biggest climatic risk to the stability of
complex societies is not global warming, but global cooling, and the
potential risks this would pose to agricultural food production. While
such a natural disaster could be triggered by large asteroid impact or
a volcanic super-eruption, the probability that any such event will
occur in any given century is remote.[30]

It would be unwise, though, to simply discard the idea of catastrophic climate change. Given the probable, albeit unproven, likelihood that climatic downturns have contributed to the decline and
even downfall of past societies, we should consider just how vulnerable our own civilisation might be to a recurring fall in temperatures.

Climatic downturns, even much-dreaded abrupt cooling events, no
longer threaten society with agricultural catastrophes, nor will they
inevitably lead to societal decline and collapse. Genetic engineering of
crops and seeds already allows the development of cold-resistant
plants. In the future, biotechnology (e.g. using a gene from a cold-resistant fish) will be able to produce cold-resistant crops that can grow
even in very cold climates. Genetic engineering has already increased
the natural defences of many crops and allowed them to survive
droughts and cold which are normally fatal to plants. There is no reason to doubt that future crops will be designed to withstand even
drier, colder and saltier conditions.

If these revolutionary developments do not reassure the alarmed
public, there has also been enormous progress in applying cloud-seed-
ing technologies to ameliorate the impact of periodic severe droughts
in a number of countries around the globe. These and other forthcoming mitigation strategies would potentially give our civilisation
technologies that could enable us to survive climatic crises which ancient and simpler societies may have found overwhelming.

Tomorrow's hyper-complex societies will be able to withstand prolonged droughts thanks to technological advances and economic efficiency. While self-reliant, agricultural societies are essentially rigid
and extremely vulnerable to climatic stress factors, inter-connected
high-technology cultures are much better sheltered from possible catastrophes, because of modern technologies and mitigation strategies.

In fact, technological progress and bio-technological developments have been advancing to levels where the age-old fears of mega-drought and mega-famine are gradually disappearing in most regions of the world. Given the accelerating evolution of disaster-resistant crops, weather engineering and other mitigation technologies, I hope that before long we will also overcome climate angst based on historical analogies that no longer match up.

11 The political economy of climate change

Carlo Stagnaro

Introduction

Scientists try to understand trends with the earth's temperatures and climate, and economists speculate about the impact of potential climate change on human activities. 'Public interest' and non-governmental organisations, especially environmental groups, use information selectively to promote certain perspectives about humanity's role and responsibility for climate change. Whether they actually represent the public interest remains to be seen, though.

The public is exposed to the views of scientists, economists and interest groups about the causes, consequences and solutions to climate change through the news media. Because their business is to sell the news, the media tend to promote the most catastrophic views about climate change. Politicians, at the national level, are driven by these groups to create political solutions to mitigate the threat of climate change, to fund further investigations on the subject, and to participate in an international regime of climate control.

This chapter suggests that European interest group politics greatly exaggerate the risks deriving from climate change and the policies needed to address it. Even in the worst scenario, global warming is a phenomenon which occurs over a long period of time. Humanity has remarkable abilities to adapt to change – and this alone could mitigate the hypothesised negative effects. Policies that are proposed to address climate change, such as the reduction of emissions through international treaties like the Kyoto Protocol, have costs which must be evaluated carefully, so that we do not spend more than the benefits we expect to receive.

The debate in Italy

The general debate on global warming and climate change in Italy is perhaps more biased than in other countries. Although there are many scientists who are reasonable about the uncertainties surrounding climate change,[1] they struggle to find voice in the news media, which gives much coverage to alarmist environmental groups and their policies.[2] For example, Grazia Francescato, former President of the Italian WWF and honorary president of the Italian Green Party, believes it is necessary to 'reduce the infamous greenhouse gases [...] not by 5.2% [...] but by 60%' to deal with the problem of global warming.[3]

There has also been little discussion in the mass media about the costs or benefits of the Kyoto Protocol. Many people believe that the Protocol is a kind of magic formula that will improve our standard of living, even though it will probably have the opposite effect.

CO_2 emissions in Italy have increased by 6% from 1990 to 1999, and it is foreseen that they will increase by a further 7% by 2010.[4] Several Italian studies have found a warming trend in this century, which is not due to the increase of maximum temperatures, but to the increase of minimum temperatures. In other words, Italy is not getting warmer, but it is getting less cold – so to speak.[5] Plants have benefited from the increased concentration of carbon dioxide,[6] and the effects of warmer temperatures were, according to scientists, not negative for agriculture.

From an economic perspective, Italy would be one of the countries most negatively affected by the Kyoto Protocol because of its industrial system, its high taxes on energy (and thus the high unit cost of emissions reductions), and its refusal to produce nuclear energy. Italy's economy is today at a difficult juncture; it faces the challenge of vast unemployment, and of a labour market that is quite rigid because of powerful unions and regulations which are inspired by a concept of 'zero risk'.[7]

It is not a coincidence that trade unions are amongst the greatest supporters of regulations inspired by environmental groups, including limitations on carbon dioxide emissions,[8] although those limitations would actually damage workers and 'capitalists' alike, and probably workers more.

For instance, the European Trade Union Confederation heartily agrees with Kyoto – probably because it would provide an inherent form of trade protectionism: 'On the whole, ETUC anticipates that the implementation of the Kyoto targets will bring positive rather than negative employment outcomes. However, it believes that the employment potential will be realised only if the appropriate political framework is set in place. In order to achieve this, ETUC calls upon governments to pursue an "active policy of green job creation".[9]

Global warming

A coalition of academics, non-governmental organisations, international agencies and governments has agreed that the primary culprit behind global warming is greenhouse gases emissions created by humans, through our use of carbon-intensive energy sources.

About 80% of the carbon dioxide created by man comes from the combustion of oil, coal and natural gas, while the remaining 20% is attributed to deforestation. However, over half of this gas is absorbed by oceans and plants. The CO_2 concentration in the atmosphere has increased by 31% from the times of pre-industrialisation – but this does not mean that such a phenomenon is due uniquely, or even mostly, to human activities.[10]

During the 1990s, the term 'greenhouse effect' became a sinister phrase associated with global warming. Of course, without the greenhouse effect, life as we know it on our planet would be impossible. The earth's atmosphere behaves similarly to the glass of a greenhouse for growing plants: it reflects part of the sun's radiation (especially ultraviolet rays), while retaining some of those that our sphere emits (especially low-frequency, high-wavelength rays – i.e., infrared rays).

In so doing, the natural greenhouse effect elevates the average temperature of the planet to about 15°C, while making thermal excursions milder. Without the greenhouse effect, the average surface temperature would be about -8°C.[11] Among the gases contributing to the greenhouse effect, the most known and important are unquestionably carbon dioxide (CO_2), water vapour and ozone.

However, Dr Robin Baker, a former reader in biology at Manchester University, describes the earth's atmosphere as 'dirty glass'. In fact, there are substances that counter warming, and even have a cooling effect. Of these, the most important are aerosols, which exist in nature as a product of marine phytoplankton, volcanic eruptions or desert sand.

The earth's climate is not simple. The reality is that many of its components can heat or cool, depending on the circumstance. Sometimes they contribute to warming the atmosphere, and sometimes to cooling it. For example, ozone shields the earth (thus making it cooler) in the stratosphere, while in the troposphere ozone works the other way around. Water vapour is a greenhouse gas, but when its concentrations exceed a certain limit, clouds are formed, and they act as if they were a mirror pointed upwards, reflecting solar radiation. In short 'water vapour's contribution to the contest is patchy, erratic and probably totally unpredictable.'[12] Climatologists and other scientists are not yet able to fully explain either the behaviour of the atmosphere, or to evaluate how individual components affect the atmosphere as a whole.

Scientists have observed an increase in average temperature of about 0.8°C starting from the middle of the nineteenth century.[13] Their measurements show that almost all the warming which has taken place in the twentieth century is concentrated in two well-defined time periods: from 1920 to 1945, and from 1975 onwards.

Humanity's carbon emissions have been rising since the Industrial Revolution, and proponents of catastrophic global warming believe that these emissions are causing global warming. But the discontinuity in observed warming in the twentieth century shows that this explanation is wrong.

The temperature variations read in the past century could be part of a larger process that is alien to humanity. A geologist at the University of Naples 'Federico II', Dr Franco Ortolani, summarises this:

> The last cold period is called the 'Small Ice Age' (1500–1850). By using new geoarcheological data it is clear that preceding cold periods have been characterised by climatic environmental conditions similar [to that one], and in fact we have defined those periods as the 'Small Ice Age of the High Middle Ages' (AD 500–750), and the 'Small Archaic Ice Age' (520–350 BC) [...] The known warm periods took place in the Middle Ages (1000–1300) and the Roman Age (AD 200–400).[14]

On the other hand, Greenland earned its name because when it was discovered by the Vikings, it was covered with vegetation. In the ninth century, vineyards were common in Scotland, while in the 1400s skating on the frozen Thames was commonplace, and malaria (often misunderstood to be solely a tropical disease) was prevalent throughout England.[15] These climatic changes are totally natural, and may be the result of a great number of causes, including solar activity, and the variations in the cycles of our star – which could also explain some of today's climatic changes.[16]

There are many reasons why systematic observations could be biased, and therefore overestimated. First, measurement stations are mostly on land, so we have relatively less information about temperature over the oceans. Second, temperature readings are often taken near urban centres, which act as heat reservoirs,[17] and these measurements may also be affected by economic and social variables.[18] Actually, satellite measurements do not seem to show significant variations in the average temperature of the atmosphere.[19]

Last, we have too little data, for all intents and purposes, to be able to comprehend the Earth's complex climate. Dr Franco Battaglia, a physicist at the Third University of Rome and former chairman of the Scientific Committee, Associazione Nazionale per la Protezione dell'Ambiente (the Italian Environmental Protection Agency) suggests that 'the only reliable data on global average temperatures concerns just the last one hundred years. It shouldn't appear strange that, if one begins [to measure] at any point in time, the global temperature either increases or decreases.'[20]

Paradoxically, therefore, we should actually be more surprised if the earth's temperature remained constant.

The news media is of particular interest with respect to scientific reporting of climate change. Newspaper headlines may announce the earth's impending demise from global warming, but the debate in the news media is one-sided and highly sensationalised. Many Europeans have been led to believe that any number of negative impacts will occur because of a changing climate – pestilence, disease, death, severe weather – even though the evidence suggests that such phenomena will not be the consequence of warming. The news media generally do not promote sensible perspectives on the scientific issues of climate change, relying instead on environmental groups for a steady stream of press releases and interesting stories.

IPCC forecasts and models

The Intergovernmental Panel on Climate Change (IPCC) was set up under the auspices of the United Nations in 1988 to study and address the issue of climate change at an international level. As such, it is one of the primary interest groups driving the debate on climate change.

To assess the potential impact of climate change, the IPCC has attempted to construct mathematical models to simulate climate behaviour from now to the year 2100. These models forecast a temperature increase between 1.5 °C and 4.5 °C. If the first estimate is true, climate change is likely to be beneficial – the earth's climate would simply be milder or moderately warmer, which could lead to substantial benefits for human health. Greater concentrations of carbon dioxide would mean that vegetation (thus crops) would grow faster and would be more abundant.[21]

But we cannot trust the IPCC's analysis on the science or economics of climate change. Its climate models are very complex mathematical tools that essentially rest on a series of simplified hypotheses. Given our lack of understanding of climatic dynamics, almost any such hypothesis is arbitrary.

So far the IPCC has not demonstrated that there is a direct link between increased greenhouse gases and increased temperature – or that the increase in atmospheric carbon dioxide is mainly (or in large degree) due to anthropogenic emissions. This is due to uncertainty about the behaviour of many substances, and some of them – such as water vapour – can have either a warming or cooling effect. Logically, a model cannot take into account such ambiguities, and thus it must be calibrated in such a way as to attribute an either positive or negative coefficient to each.

The IPCC's models have always assumed a positive feedback among all the substances that contribute to global warming. Particularly, it has been assumed that more carbon dioxide warms the earth and leads to more water vapour being released into the atmosphere, which in turn would further contribute to temperature increase. However, when water vapour creates clouds, this actually cools the atmosphere, so temperatures are pushed in the opposite direction. We don't know if the latter contribution is greater than the former, but if it is, then all of the IPCC's models have a tendency to overestimate temperature increases by a great deal.

As Robin Baker states in *Fragile Science: The Reality Behind the Headlines*:

if the positive feedback on which computer models depend is largely imaginary, the climate could simply be less sensitive to human activity than we were first warned. *The only thing that is certain is that the current models are neither powerful enough, sophisticated enough, or informed enough to be able to decide.*[22]

The IPCC's economic forecasts are equally unreliable, as they rest on a foundation of assumptions about a warming climate. In a letter addressed to Rajaendra Pachauri, chairman of the IPCC, Australian statistician Ian Castles alleged that there are many 'wrong hypotheses', concerning in particular the economic growth of less-developed countries. Dr Castles wrote:

I believe that it is important that governments be advised as soon as possible that the economic projections used in the IPCC emissions scenarios are technically unsound, having been derived by converting national GDPs in nominal values into a common currency using exchange rates. This procedure is not permissible under the internationally recognised *System of National Accounts,* and was recently rejected by an expert group in a report to the UN Statistical Commission. The practice of using exchange rate conversion is especially inappropriate in relation to projections of physical phenomena such as emissions of greenhouse gases and aerosols.[23]

Many of the IPCC's economic forecasts treat global warming as if it would occur in a very short period of time, and not gradually throughout the course of an entire century. As Brookings Institution's Robert Crandall observed:

Regardless of the model used, all forecasts of global warming see only a gradual warming over the next few decades or centuries. The alleged problems from the delayed impact of past and future greenhouse-gas accumulations do not become serious for at least fifty or sixty years. Every dollar dedicated to greenhouse-gas abatement *today* could be invested to grow into $150 in the next 50 years at a ten percent social rate of return, and even at a puny five percent annual return, each dollar would grow into $12 in 50 years. Therefore, we need to be sure that the prospective benefits, when realized, are at least 12 to 150 times the current cost of securing them. Otherwise, we should simply not act, but use our scarce resources in other ways.[24]

Lastly, to evaluate the opportunity to limit warming (assuming that it is primarily caused by humanity's emissions of carbon dioxide) it is necessary to understand the costs to humanity if we choose to do business as usual. 'The total annual cost of all the considered global warming problems is estimated to be around 1.5–2 percent of the current global GDP, i.e. between 480 and 640 billion dollars.'[25] This figure should be approached with hesitation for the reasons that have just been explained, but it may act as a yardstick for comparison.

Adaptation

The interest groups driving the climate change debate promote the view that politicians must act quickly, because climate change is imminent and its negative effects are already occurring. Climate change is viewed as a problem in itself – environmentalists and others argue that we must 'do something' urgently, regardless of the costs or benefits of that action.

However, global warming is a problem only if it presents a danger to the well-being or survival of humanity. Changes in the earth's climate will most certainly happen, but these will occur over the long run, and we do have time to rationally consider any number of potential responses. It is of utmost importance to focus on the effects that climate change would have on poor and rich countries alike, and how we can adapt to such changes.

Humanity has adapted to change (climatic or otherwise) through technology, and through markets. During our evolution as a species and as civilisations, humans have modified the environment, first with agriculture, and developed more efficient technologies to feed, clothe and shelter ourselves, to be transported from place to place, and to improve the well-being of many people.

Without adapting our environment, we would be much less able to protect it. For instance, we would need to plough under larger areas of land, entailing deforestation and loss of biodiversity, to feed the earth's 6+ billion population. Between 1950 and 1989, although the world's population doubled, the land used for agricultural purposes grew only by 26%. Per capita land surface utilisation has decreased from 0.45 to 0.28 hectares, while per capita food consumption has increased everywhere, in both wealthy and poor countries.[26] A similar phenomenon has been observed for every other challenge presented to humanity.

Economist Julian Simon emphasised that our 'ultimate resource' is human intelligence, which is expressed through our minds, our creativity, and our ability to address and solve problems in an original manner, thus creating a better world for future generations. Without the need to warm themselves, our ancestors would have not discovered fire; and if that had not happened, we might still live in caves.

Individual efforts to solve particular problems, in the form of new technologies, are harnessed by markets, which leave humanity better off in the long run.[27] New technologies supply the means to obtain better goods and services with fewer resources, fewer negative environmental consequences, and at a lower economic cost. For example, today's car engines cost, burn and pollute far less than those of past decades. By the same token, energy sources such as carbon-intensive fuels will be gradually replaced with cleaner and more efficient alternatives.[28]

In the long run, economic growth results in a cleaner environment, because wealthier societies generally can afford to shift their priorities from mere day-to-day survival to aesthetic concerns.

Free markets, unhindered by subsidies or trade barriers, are fundamental to creating economic growth. Markets harness new technologies, stimulate the circulation of ideas, information, goods and services. They create a closed loop of economic interdependence and labour skills which, in turn, produces wealth and welfare. As analyst Indur Goklany observes:

> Even in the absence of formalized agreements on technology transfer, throughout history, trade has helped disseminate both 'hardware' and 'software' technology, i.e., knowledge, ideas and modes of thinking around the world, some of which have helped shape our political, legal and economic frameworks.[29]

(See also Chapter 3 'Climate change: the 21st century's most urgent environmental problem or proverbial last straw?')

The mere opportunity to engage in economic exchange has always allowed societies to improve their well-being. On the other hand, closed systems and planned economies have damaged not just human flourishing, but also the environment.[30]

A framework for adaptation may entail eliminating some of the rent-seeking ability of interest groups, for instance, removing subsidies for all forms of energy (including renewables), encouraging technological innovations and adopting those technologies, and eliminating regulatory barriers which stifle economic activity and distort prices. While this might be a difficult short-run strategy, in the long run it would yield greater benefits.

Certainly, risks presented by global warming should not be underestimated – but they shouldn't be overestimated, either. It is a problem, not a tragedy, and we must unleash human creativity to adapt to it.

The costs of Kyoto

The interest groups driving the Kyoto Protocol in Europe have failed to illustrate to the public that pursuing a mitigation policy is not without cost. Of course, this is in their self interest. Mostly, the debate has focused on the urgent need to react to climate change, without a careful consideration of the costs and benefits of various strategies.

The Kyoto Protocol is an agreement which forms part of the United Nations Framework Convention on Climate Change. It mandates that countries which have ratified the treaty will reduce their carbon dioxide emissions by precise and significant amounts. The Protocol focuses on limiting greenhouse gas emissions, but it doesn't address what is considered to be the actual problem, which is atmospheric concentrations of greenhouse gases.[31]

Under Kyoto, European Union countries committed to reducing their emissions by 8% under their emissions in 1990, and some committed to even stricter targets.[32] Poor countries are excluded from Kyoto, although they contribute to about 50% of worldwide emissions. By 2050, that figure may rise to 75% of global emissions.[33] On the other hand, asking poor countries to adopt reductions similar to wealthy countries would have devastating effects on their economies and economic growth.[34]

If the countries listed in the Protocol's Annex I fully accepted and enforced its requests, the warming would be minimally mitigated, that

212 Adapt or Die

is, by only 3–10% in a one-century period.[35] Bjørn Lomborg, author of *The Skeptical Environmentalist*, notes that Kyoto will do little to help the earth's climate, with 'a temperature increase by 2100 of around 0.15 °C less than if nothing had be done', and this would occur six years later – in 2100 rather than in 2094.[36]

So the Kyoto Protocol is not enough to stave off climate change. If we want to act seriously against human-caused global warming, Kyoto is only a first step towards a crackdown that would be much more severe, and would involve every country in the world.

A super-Kyoto regime implies at least two consequences of great importance, neither of which has been highlighted by interest groups in Europe.

First, the use of energy for food production, refrigeration, transportation, heating, manufacturing and air conditioning would be greatly curtailed. Affordable, reliable energy has enabled human beings to live longer, healthier, happier lives. People, especially Europeans, would be forced to greatly curtail or give up their use of energy, leading to a drastic reduction in quality of life.

Second, a 'super-Kyoto' would entail a global enforcement mechanism, through central planning by global agencies such as the United Nations, a prospect viewed with suspicion by many people. Poor countries would likely see this as a kind of 'ecological imperialism' against their desire to obtain a better quality of life through economic growth, which relies on more intensive energy use.

The impact of the Protocol would be at any rate differentiated even within the OECD countries. The main factors that determine the effect of Kyoto measures are:

- The rate of emissions growth, which is affected by variables such as demographic trends, income growth, and the relative importance of the various energy sources. Greater emissions reductions imply greater loss in GNP terms.
- Where already high taxation exists on energy (i.e. in Italy, France and Japan) it is necessary to impose even heavier taxation to obtain reductions in emissions.
- The countries that not only have important industrial sectors that are based on an intensive use of energy, but also participate in vast trade with other countries – especially the OECD ones – face greater sacrifices.[37]

Countries that have economies based on an intensive use of energy would be subject to greater costs.

France and Italy are the European countries that must sustain the largest expenditures because of the Protocol. France, in particular, produces about one-third of its energy with nuclear power. France will have to reduce emissions by a smaller amount, but the unit price of those reductions will be greater. Germany and the UK are the countries less affected by Kyoto, especially because of the slow growth of their economies.

The UK can be considered as a 'minimum threshold' for European countries. According to a study performed by DRI–WEFA, the price of oil for heating purposes in that country would increase by 46%; diesel and petrol by 13% and 10% respectively; industry would pay about 117% more for natural gas; and the cost of energy in general would double.[38] The most critical period would be between 2008 and 2012. The GNP would be reduced by up to 4.5%, and it would not return to normal levels until 2020.

For the same reasons, economic production would decrease because of increased energy costs. Moreover, a decrease in consumption could cause a depression in the short term. Between 2008 and 2010, the UK risks losing up to one million jobs each year. The productivity of individual jobs would decrease because of the efficiency reduction (greater cost) of all the other production factors.[39]

In May 2003, the European Environment Agency announced that in 2002 most EU countries did not reach their emissions targets under the Kyoto Protocol.[40] The UK has already contributed heavily to initial reductions of emissions, by substituting natural gas for coal, as has Germany, which renewed the inefficient industry of East Germany after reunification. On the other hand, ten of the fifteen member countries increased their emissions during the 1990s.

According to the projections, Europe's emissions in general will increase by 9% by 2020, unless dramatic political choices are undertaken to curb them. The DRI–WEFA study was extended to the entire EU, and it estimated that such a policy would have a strong impact on the GNP of various nations: a decrease of 5.2% for Germany, 5% for Spain, 4.5% for the UK, and 3.8% for the Netherlands.[41]

What would this mean for the average European citizen? Consumers would see rapid increases in living costs – food, durable goods, heating and cooling, transportation – because all energy, not just oil and gas, would be more expensive. If emissions limits were established, the cost would be passed on by businesses to consumers. Combined with the increased cost of energy, consumers would see the buying power of their salaries greatly weakened. Because economic production would greatly slow, many people might lose their jobs.

The cost of the Kyoto Protocol for the EU is very steep indeed. Its benefits – the actual impact on the earth's climate that its emissions cuts would achieve – are negligible. Europeans should ask themselves whether they are willing to give up their well-being – economic growth, the buying power of their salaries, and many jobs – in exchange for a hypothetical slowdown of global warming.

Market solutions ... or markets?

It is alleged by some economists that environmental problems such as global warming are 'market failures'. Specifically, they mean that markets result in 'externalities' – unintended and/or undesirable effects for third parties. They often propose that government intervention is required to solve such problems – usually in the form of regulations, subsidies, and by generally limiting human activities.

Using the tools of public choice analysis, global warming policy is no different than other public policies. The political arena naturally produces inefficient rules that are designed for vested interests such as economic pressure groups, bureaucrats, and the political elite.[42]

Some economists, including those at the IPCC, do realise that markets achieve allocation of scarce resources better than governments do. They have recommended a 'market mechanism' (which is included in the Kyoto Protocol) to deal with the problem of CO_2 emissions, through emissions trading schemes where governments allocate a certain number of emissions permits, which are then traded by companies and valued by the market. This scheme is now being tried in Europe.

One problem with such market mechanisms is the allocation of emissions permits.

An alternative approach, based on markets rather than market mechanisms, is to treat carbon dioxide as a common law pollutant, if it is established that carbon dioxide emissions are excessive, and they are causing harm to third parties.[43] Pollution implies that a right has been violated, and that violation should be compensated through a legal remedy. Rather than encouraging rent-seeking by interest groups, the common law protects the right not to be polluted upon.[44]

If this is the right way to approach global warming, then quotas should not be viewed as a panacea. Indeed, they are part of the 'myth of efficient central planning', as named by Dr Roy Cordato.[45] The main strike against the permit system is that governments cannot possess information which is spread amongst millions and millions of actors in the market. Order evolves in this context from competition and independent actions, but with central planning it does not. Thus, central planning can never be efficient or produce the best allocation of resources.

The Kyoto Protocol's justification is also quite controversial. Interest groups suggest that we must decide today to pursue expensive climate mitigation strategies which will produce a net benefit for future generations. This assumes that we know a whole series of preferences that logically and practically cannot be known: neither by people living now nor by future generations.[46] Thus, we cannot know the costs of our choices today, so we are simply making blind choices.[47]

We live in a world characterised by a scarcity of resources, including human resources. The price of any resource reflects its relative scarcity – and through markets, prices transmit information to other economic actors. Just as it fails to allocate resources efficiently, government's intervention generally prevents a discovery process to solve pollution-related problems. In the long run, this prevents creative individuals from finding a solution for the problem.[48]

Moreover, if we view the greenhouse effect as a possible cause of external costs, then we should not view it as completely catastrophic. Instead, we should find ways to assign a price to that pollution – then private actors can flexibly find ways to minimise their pollution, and they have an incentive to develop new technologies and substitutes. But environmental groups are often completely unrealistic about this, on the one hand painting scary scenarios of resource scarcity, and on the other hand demanding the taxpayer-subsidised adoption of highly expensive energy sources.[49]

Conclusions

We should be concerned about the earth's future, but too much worrying can be unproductive. It is important to get our priorities straight. So far in Europe the debate on climate change has not in-cluded the scientific uncertainties which surround the debate.

Scientists do not fully understand climate dynamics, and climate models are not sufficiently reliable. Amongst a variety of other uncer-tainties, there is a discrepancy between the earth's surface tempera-tures, which show warming, and satellite readings, which do not show evidence of warming.

For the same reasons, economic forecasts on the impact of global warming or of the Kyoto Protocol must be regarded with scepticism, because often they are based on excessive simplifications or on unreli-able information. We also do not fully comprehend the potential ef-fects of climate change for humanity or the environment. These could be positive or negative, but we should be certain before we act to mit-igate climate change.

The European debate has also lacked a discussion of how we should prioritise climate change, amongst other policy priorities. It has been one-sided and driven by interest groups with a variety of agendas, some of whom believe it is our moral duty to act urgently against climate change, and others of whom stand to gain from rent-seeking relating to the Kyoto Protocol and supporting regulations.

Bruce Yandle, an economist at Clemson University, suggests that:

> An analysis of the agreement and of the post-Kyoto strategizing
> suggests that control of global warming is largely symbolic, which
> does not gainsay its vital importance to environmental groups.
> The real effects of the Protocol relate to cartelization and efforts
> by interest groups and countries to gain competitive advantage in
> a globally competitive world. Global warming may be just the
> right wrapping for a major rent-seeking package.[50]

(See also Chapter 9, ' Bootleggers, Baptists and the global warming battle'.)

Europe's policies should be driven by an analysis of their costs and benefits – not by reliance on arbitrary criteria such as the 'precaution-ary principle'. Kyoto will produce marginal effects in avoiding global warming, but its costs will be extremely high, for wealthy and poor countries, as outlined in this chapter.

More importantly, Europeans should encourage and adopt policies that allow flexibility, innovation and adaptation to change – whatever those changes may be. By rationally facing potential problems, we will avoid wasting and diverting resource from more urgent and substantial needs. To that end, we should:

- Invest in climate research, both scientific and economic. A better understanding of the process and its effects is indispensable to creating public policy.
- Encourage innovation in energy sources and utilisation.
- Stimulate adaptation to change.
- Remove obstacles to the scientific process and to economic growth, including subsidies and trade barriers.

To achieve these results, it is necessary to:

- Guarantee freedom of research for scientists.
- Strengthen the institutions of free societies, to enable human innovation and creativity.
- Support free market reforms which are unbiased towards specific energy sources.

Epilogue

Kendra Okonski

More than just a problem, the threat of global warming offers an important opportunity – at the beginning of the third millennium – to evaluate the progress, successes and failures of humanity thus far, to contemplate where we are going as a civilisation, and what sort of world, environmentally, culturally and economically, we would like future generations to inherit from us.

But concern for the future need not, and indeed should not, entail neglecting the present. Despite the efforts of climate campaigners to convince us otherwise, the biggest challenge facing humanity today is the elimination of poverty. A minority of people on our lovely blue-and-green planet lead long, prosperous, happy lives. Others, a minimum of at least 2 billion people, lead lives of extreme poverty – their lives are akin to those of our ancestors, defined by drudgery, misery, disease and early death.

As the global warming debate has progressed since the early 1990s, it has become increasingly clear that two possible paths could be followed to address the impacts of a warmer climate.

The 'climate control' approach dogmatically suggests that the threat of global warming looms so large that we must control the earth's climate. According to this view, the wealth of some countries has caused poverty in others, and the well-being of some (the minority) negates the well-being of the majority. It is claimed that environmental problems, including global warming, have been caused by reckless consumption of resources. Poor countries, it is argued, must follow a different developmental path. 'Climate control' also means that wealthy countries must urgently reduce their consumption of energy and resources. Those who favour climate control hold, as an article of faith, that such sacrifices are the minimum necessary to prevent catastrophe.

The other approach recognises that humanity plays a role in the earth's climate, and that some adversity might result in the future because of global warming. However, there are many options available to humanity to address potential threats (whether or not they result from climatic change) which do not entail climate control. This strategy, called 'adaptation', relies on institutions to encourage humanity to adapt to change and to address potential threats, whether they occur in the year 2100, 2150 or 3500. Such a framework is the best way to achieve human *and* environmental well-being, and sustainability.

The climate debate thus far has lacked coherent discussion of how adaptation measures could help humanity to address the potential impacts of a warmer climate, and this book is an effort to foster such discussion. Amidst many different scenarios, how can we respond to potential global problems, and how can we encourage better decision-making to achieve both our present and future priorities?

The climate debate in the 1990s

In the mid-nineteenth century, a French scientist called Jean Baptiste-Joseph Fourier suggested that the earth's atmosphere acts to trap the sun's heat. Indeed, without this 'greenhouse effect' the earth would be uninhabitable. Since Fourier's time, and especially in the past three decades, thousands of scientists have investigated the earth's climate and have speculated about how humanity's activities might contribute to this complex system.

But it was only recently that the phenomenon of 'global warming' entered people's radar screens. Indeed, in *Green Warriors: The People and Politics Behind the Environmental Revolution*, Fred Pearce shows that environmental groups 'were very late to wake up to the greenhouse threat'. Though scientists had been warning for several decades 'that carbon dioxide was building up in the atmosphere and could be expected, within a few decades, to start warming ... Somehow none of the thrusting campaigners felt confident enough either about the science or the public's interest to take up the issue.'[1]

This changed in 1988 when, 'at the height of an unprecedented drought across the US', James Hansen, a NASA scientist in the USA, 'got the greenhouse effect onto [the front pages of newspapers]' by claiming that the drought was caused by global warming. 'Once alerted, and with journalists calling for their views, the greens were full of warnings, admonitions and quotations. Some spoke as if they had been in this greenhouse business for years.'[2]

Pearce suggests that the greenhouse effect offered an opportunity to environmental campaigners who were 'looking for new challenges'.[3] They latched on to the issue of climate change, promoting a series of apocalyptic claims which injected the debate with a sense of urgency and alarmism.

In 1990, the Intergovernmental Panel on Climate Change – a joint effort between the World Meteorological Organization and the United Nations (UN) Environment Programme – issued its First Assessment Report on the state of the earth's climate. Subsequently, in 1992 governments and bureaucrats, environmental groups and an array of other characters met in Rio de Janeiro, Brazil, for the UN Conference on Environment and Development, also known as the 'Earth Summit'.

It was during the Earth Summit that the UN's Framework Convention on Climate Change (FCCC) was formulated to deal with the problem of global warming. From 1992 onwards the debate shifted to implementation of the FCCC, which took for granted that reductions in greenhouse gas emissions would be the basic tool for climate control. After 1992, signatories to the IPCC negotiated a Protocol that would specify explicit reductions in emissions of GHGs. The Protocol was agreed in Kyoto in December 1997, and has subsequently been ratified by 118 countries. However, it has not yet entered into force because too few 'Annex 1' countries have ratified. To enter into force, Australia, the USA or Russia would have to ratify.

NGOs, and especially environmental groups, have played a critical role in popularising the idea of global warming, in particular by sensationalising the effects of a warmer climate. For instance, Friends of the Earth calls global warming 'a terrifying prospect'.[4]

Most people are rationally ignorant of the real causes and consequences of a phenomenon such as global warming. They make decisions about how such issues might affect their immediate well-being and how they might confirm or conflict with their underlying values – such as protecting the environment or achieving a fairer society. Europeans believe that environmental groups have the interest of the earth at heart, so they are trusted purveyors of information.

It is for this reason that environmental campaigners focused on the impacts of a changing climate. Vivid images provided by hurricanes, floods, drought and disease caused by a warmer climate have resonated well with members of the public. Thus global warming has become conventional wisdom.

But the spectre of such negative consequences also enabled environmental campaigners to convince people that our profligate use of energy and consumption of resources in wealthy countries is the cause of global warming, and that sacrifices are needed in order to solve it.

The debate about globalisation has also relied on environmental problems such as climate change to justify environmental campaigners' calls for less trade. Trade is deemed to accelerate global warming because economic exchange requires production, manufacturing, and transportation based on hydrocarbon fuels. The International Forum on Globalization, for instance, alleges that 'globalization accelerates warming-caused disasters by dramatically expanding industrial activity and universalizing the carbon-intensive model of development worldwide'[5] and 'no domain of global economic activity does greater social, environmental and political harm than presently dominant energy systems ...'[6]

The campaigners' suggestions to reverse the course of global warming have focused on attempting to control the earth's climate, by reducing humanity's emissions of greenhouse gases. They constantly remind us that concentrations of carbon dioxide in the earth's atmosphere are far higher than before the Industrial Revolution. This buildup has been caused by 'reckless consumption' of scarce resources, which has resulted in unnecessary emissions of GHGs.

This results from the idea that traditional economic growth is unsustainable, and that consumption should be made more costly and thus less desirable. The Kyoto Protocol is one such effort. To the anti-globalists, climate change is just one manifestation of mankind's evil obsession with consumption. The spectre of fireball earth presents them with a perfect vehicle for declaring the need to make our actions more sustainable. While environmental campaigners concede that Kyoto itself may do little to change the earth's temperatures over the next 100 years, they argue that it is an important first step towards a broader goal of reducing economic activity by reducing global emissions of greenhouse gases through 'contraction and convergence':

the proposal that the total amount of emissions produced globally should contract over the next few decades by the IPCC's recommendation of at least a 60 percent reduction. As emissions steadily reduced, the more industrialized countries that would still have higher emissions would have to buy quotas from the countries that emitted less.[7]

Friends of the Earth agrees with this goal: 'globally we need to make at least a 50 percent cut in CO_2 emissions to stabilize the climate. In the UK we need a 90 percent cut to allow for fair shares across the globe.'[8] This claim is problematic on many levels, not least because it presumes, counterfactually, that GHGs are the primary driver of climate change.

In early 2003, the UK government released an Energy White Paper, which states similar goals for the UK. By 2050, it hopes to reduce the use of hydrocarbon energy to achieve the target of reducing carbon dioxide emissions by 60%.

'Climate control' strategies are often justified by the claim that people today have a moral obligation to protect the earth for future generations and must alter their behaviour, and especially their consumption, so as not to compromise the ability of future human beings 'to meet their own needs'.

The underlying premise of the climate controllers is the idea that resources must be shared equally between generations. Consumption by any individual today entails less of something for someone else to consume, whether today or tomorrow. Thus, wealth which is generated today means poverty for someone else tomorrow.

The idea may sound appealing, but as a guiding principle for action today, it fails to account for the benefits that our activities and decisions may have for future generations. In an editorial last year in the UK's *Sunday Telegraph* at the time of the 2002 World Summit on Sustainable Development, a South African economist suggested that 'The best we can do for future generations is generate maximal wealth to let them live better lives ... If anything is unsustainable it is the alternatives to development: stagnation and regression.'[9]

If an obligation to future generations is taken as an action-guiding principle, the implications are absurd. There is a possibility that *any* decision we make today may have an unforeseen negative consequence on people in the future. Should we therefore eliminate any such actions?

Even with the best intentions, shifting our focus to future generations would cause misguided priorities and confusion. Some courses
of action do make sense when considering the lives of future people,
but many do not. Moreover, it is patronising. People everywhere plan
for their futures, often in spite of failed central planning by bureaucrats, and generally they do this quite well.

By focusing our priorities on 'future generations', we focus less on
improving the lives of people who are alive today. Development for
people today – not just meeting people's needs, but allowing them to
achieve real well-being – means that people today are better off, and
future generations will benefit from our knowledge, investment, and
even our mistakes, which for the most part are part of an ongoing discovery process and not part of some global catastrophe.

Trade-offs and priorities

Some academics and environmental campaigners suggest that global
warming is such an urgent threat to humanity and future generations
that climate control – in the form of reducing emissions, consuming
fewer resources, engaging in less trade – should take precedence over
any other course of action. They stress the urgency of the problem,
claiming that the possible consequences of global warming are so dire
that all immediate costs and actions are justified.

This book has shown that some of the possible impacts of warming, such as biodiversity losses, shifts in agricultural production, the
spread of disease, extreme weather and sea-level rise. The reality is
that while many such problems exist, they may not be caused by
warmer temperatures. If they are unabated, such problems could be
exacerbated by a warmer climate, but this is an argument for controlling those problems, not necessarily for controlling the climate.

The question 'What is the best way to address these impacts?' has
seldom been posed.

Because we do not live in an ideal world, policies have trade-offs.
Whatever policy course we pursue necessarily has costs (even 'doing
nothing' has costs) – including explicit costs, such as enforcement and
implementation, and also implicit costs, such as the benefits we forgo
by not pursuing other courses of action.

Trade-offs in the climate change debate are illustrated well by the
controversy in the historical city of Venice, highlighted by Dominic

Standish. The city is sinking, and the problem may be made worse by rising sea levels. But Project MOSE, a system of mobile barriers which would minimise the effects of one problem (rising sea levels) has been vehemently opposed by environmental campaigners in Italy. Standish explains

> Those with a preservationist mentality – such as the opponents of Project MOSE – believe that we can have a world without trade-offs. According to this mindset, intervening in nature for human benefit is inconceivable as it [would] change the state of the lagoon.

Still, many environmental activists and academics believe that trade-offs and costs should not be a consideration in the policies that we pursue. This view presumes that the earth should not be viewed in such crass terms as 'cost', nor should decisions about the earth – especially global common resources – be based on cost. Also, it implies that the well-being of humanity must play second fiddle to the protection of the planet.

This logic implies that the earth's value is infinite, and thus an infinite number of costs are justified to protect it. But in the real world, individuals must use some metric to decide how to use their scarce time, intelligence, money and other resources in myriad number of ways which fit individuals' goals and desires. The preferred metric of many such decisions is financial cost. This is no less true with our approach to environmental issues. There are many paths to sustainability, but to decide which one to follow we must establish priorities.

If the priority is to protect ourselves from the potential impacts of a warmer climate, Indur Goklany (Chapter 3) suggests that we can achieve this by

> reducing today's urgent public health and environmental threats (such as malaria, water stress, hunger and habitat loss) that might be exacerbated by climate change. This would provide greater, more cost-effective and quicker benefits to both humanity and the rest of nature.

Indeed, as Goklany suggests, actions taken to increase resilience and reduce vulnerability may even raise the threshold at which we need to stabilise greenhouse gas emissions in the future.

The impacts of a warmer climate may adversely affect people in both wealthy and poor countries, but poor countries could be more vulnerable to those impacts. But is climate control the best way to help poor people?

The climate change debate in particular has promoted the idea that wealth has caused the world's environmental problems, and that the wealth of some has caused the poverty of others, because they consume a disproportionate amount of natural resources. According to one group, 'At the rate humans misuse the Earth we are already causing global ecological and climatic disruption. To provide a decent lifestyle for all the world's people in the coming century in the wasteful way we do in Britain today would require the resources of eight planets – yet we have just one.'[10]

This logic would suggest that everyone – but particularly people in wealthy countries – must consume fewer resources to avoid further ecological and climatic disruption, to 'save the earth'. Likewise, it would be unacceptable for people in poor countries to achieve the same living standards as those of wealthy people today, for this may lead to similar problems. Resource use must be 'fair', defined at some arbitrary level. Sacrifices are required because environmental apocalypse is nigh.

The evidence flies in the face of such myths. As Indur Goklany notes, 'almost every indicator of human or environmental well-being improves with wealth' (see Figure 4, p. 68) – people live longer, are better nourished, have lower mortality rates, have better access to clean water and sanitation, education and water, and have cleaner environments.

The reality is that wealth allows humanity to lead a more environmentally benign existence. Wealth is created because, as Peruvian economist Hernando de Soto has shown, people are driven by a desire to engage in economic activity and make better use of their resources. They innovate new technologies, which allow us to protect resources, while using fewer of others. In essence, this wealth allows us to make more out of less, to utilise new resources and to cease using others. Whales (which were formerly utilised for their oil) were saved because of the advent of hydrocarbon energy, including petroleum. The invention of fibreoptic cables, which replaced copper wires, brought about the advent of our current electronic age, with corresponding gains in time and savings in resources. Modern agriculture has saved millions of acres of wild, biodiverse land while feeding people better.

But problems remain, and they are mostly due to poverty. Indoor air pollution, caused by the burning of traditional energies such as wood and dung, contributes to over 2 million deaths each year (particularly from acute lower respiratory infections in children), and dirty water and poor sanitation are one of the leading causes of premature deaths. Several million children die each year from preventable diseases which result from poverty. Low-yield agriculture encourages the conversion of wild lands into farmland. A lack of clean and affordable energy in poor countries is causing deforestation, erosion and loss of biodiversity. Lack of electrification, refrigeration and packaging means that a large quantity of food spoils before it reaches consumers.

Development is not just about fulfilling poor people's basic needs, but allowing them to choose how they develop and to choose which technologies they use.

While such indicators do not guarantee human happiness, they can be considered a critical measure of humans' well-being – if people's basic needs are met, they are more likely to achieve other aspirations beyond survival.

It is fundamentally wrong, in fact backward, to think that the wealth of some must come only through the poverty of others. It is a fallacy of composition based on the misconceived notion that economic exchange – trade – is a zero-sum game.

If we took seriously the notion that we need 'eight earths' to meet our needs, no doubt we would end up preventing all sorts of desirable and efficiency-enhancing, environment-protecting activities, in the name of consuming fewer resources.

Access to new technology will allow poor people to use their resources more efficiently, to be healthier and to live a more benign existence. Such technologies are not an end in themselves – they allow people to work fewer hours and with less effort, to earn a living rather than subsist, to control their environment and to invest in the future of their children, their community and their country, as well as their environment.

As Barun Mitra (Chapter 5) notes, this is a 'virtuous cycle of development'. For instance,

Cleaner, more efficient energies would lead to less degradation and more efficient use of resources. Ultimately, this would lead to conservation and efficiency in the energy sector. Wealthy countries have illustrated that this path of development works, and poor countries should not be discouraged from following it as well.

But the likely result of international climate policy would be to slow the development of poor countries, by discouraging investment in new energy sources, and by making energy technologies less affordable.

Poor countries and global warming

Much of the global warming debate has focused on the potential vulnerabilities of poor countries to the effects of a warmer climate.

Bangladesh, for instance, is a low-lying country which is the delta for three rivers, and is inhabited by 133 million people, most of whom live in extreme poverty, often in makeshift shacks. Bangladesh has constantly been at risk of storms and flooding, which frequently cause the displacement of many people as well as devastating their economic livelihoods. Many have advocated 'climate control' to solve such problems.

Small island states have also claimed that governments should urgently adopt the Kyoto Protocol because they may be inundated. These have banded together to lobby for Kyoto as the 'Alliance of Small Island States'.

But if flooding or sea-level rise could be exacerbated by a warmer climate, this offers an excellent opportunity to consider how poor countries can improve resilience to such problems.

A Sri Lankan newspaper recently opined that:

> Low-cost, flood-resistant housing designs, promoted by
> organisations like the Intermediate Technology Development
> Group (ITDG) in countries like Bangladesh are simple yet
> versatile. Some homes are built on concrete stilts and have a
> roughly made attic, where family possessions and food can be
> stored when the water level rises. Many of these designs are meant
> for extremely poor, one-room homes. But the general idea of
> constructing flood-proof houses should be adopted in Sri Lanka
> and adapted to suit the location and socio-economic status of the
> wet-zone populace.[12]

Of course, a longer-term solution would be to examine what factors are preventing poor people from escaping poverty, rather than focusing on controlling the earth's climate.

Poor countries generally stay poor because they lack institutions which support development. Both Bangladesh and Nigeria, like many other poor countries, are politically unstable, ruled by a small elite (especially concerning resources such as oil), and a large percentage of economic activity is informal. They lack adequate formal legal institutions such as property rights and the ability to form contracts – which provide the impetus for development.

How does business promote sustainability?

While consumers are made to feel guilty about their consumption, environmental campaigners believe that the real cause is businesses. In particular, businesses which requires energy (nearly all) – and businesses which have enabled more people to consume items such as refrigerators, cars and air conditioners – have been lambasted by governments and environmental campaigners in the global warming debate.

But some large businesses have responded by becoming intimately involved with climate policy negotiations, so that climate policy would be designed to benefit their interests. As Bruce Yandle and Stuart Buck show, 'government subsidies have created opportunities for rent-seeking on both sides of the global warming controversy.' If government can pick winners and losers, it is logical to expect that 'tax breaks, subsidies, and regulations requiring energy-efficient products are sure to be supported by the manufacturers of those products.'

Of course, small and medium-sized European businesses (SMEs) have had very little, if any, input into this process. SMEs provide two-thirds of employment – about 80 million people in Europe.[13] It is these businesses and their employees who have suffered at the hand of European climate policy, and will continue to do so, since they do not have sufficient resources to affect the debate.

Businesses have made it possible for average people in many countries to access new technologies. They have done this by developing new products, and finding more efficient ways to produce goods and services, thereby making those items affordable to more consumers.

Business has made our environmental footprint lighter, because it constantly finds new ways to utilise resources more efficiently. Businesses have helped to make modern agricultural technologies widely available – especially during the twentieth century – so that we produce more food on less land.

People who work for businesses do not awake in the morning to the thought of 'How can I best pollute the planet today?' As Martin Livermore (Chapter 8) notes:

> Businesses do not set out with the intention to pollute the environment, use up natural resources, or encourage profligate consumption. Businesses naturally strive for improved efficiency and profitability – and generally, this means that their activities are more environmentally benign.

Businesses of all sizes will play a key role in our adaptation to climate change. Business drives resource efficiency – and ultimately businesses will find ways to make new technologies affordable to consumers everywhere, leading us into a sustainable future.

Some environmental campaigners may continue to be deeply cynical about business, blaming it for all manner of alleged evils and attempting to foist global rules for corporate accountability upon it. Businesses should ignore this cynicism, and focus on improving human well-being by helping us to adapt to change.

The evolution of *Homo sapiens* over several million years illustrates that humans utilise change rather than viewing it as a catastrophe. Specifically, we have applied our intelligence to challenges in order to survive. Adaptation has allowed us to survive change through modifying our environment, using our resources more efficiently, dealing with potential harms and threats, and improving our understanding of the world around us. The dawn of fire, the wheel, agricultural cultivation and the Industrial Revolution indicate that humans have adapted successfully to change.

Adaptation, as defined by Indur Goklany (Chapter 3) is

> measures, approaches or strategies that would help cope with, take advantage of, or reduce vulnerability to the impacts of global warming.

Martin Ågerup (Chapter 4) suggests that adaptation is inherently human-centred and dynamic:

> With the right incentives and institutional framework, people
> solve problems, they rise to challenges and adapt. They invent new
> technologies and new ways of doing things. And over the course
> of an entire century they will have plenty of time to do so. Some of
> the climate change models assume that people don't adapt to new
> conditions – that is not a reasonable assumption.

The following suggestions are some of the features of an appropriate institutional framework to encourage adaptation.

1 **Decentralise responsibility** by devolving both incentives and
   responsibility for decision-making to the lowest possible
   level, be that individuals, businesses, communities. People in
   poor countries today have suffered at the hand of
   bureaucrats and the elite, who have used laws to stifle
   development.
      Institutions such as property rights and contracts provide
   incentives to adapt, because the benefits of adaptation accrue
   to those who invest their time, effort and resources in pursuit
   of better solutions. These institutions also ensure that
   responsibility for the outcome, whether good or bad, lies with
   the decision maker, rather than with third parties.
2 **Foster an environment of certainty.** Property rights and
   contracts require a system based on the institution of the rule
   of law. The rule of law ensures that property rights and
   contracts can be defined, enforced, and transferred, which
   affords an environment of certainty in which to make
   decisions.
3 **Encourage flexibility.** Institutions for adaptation should
   tolerate human errors, which are inevitable. It should
   encourage us to learn from those mistakes but also ensure that,
   largely, people are not harmed or they are compensated if
   harmed.
4 **Ensure that benefits are widely shared.** Institutions such as
   open trade, freedom of movement and regulations based on
   rational risk assessments and sound science encourage both the

development of new technologies and the sharing of those benefits amongst humanity.

5 **Focus on processes rather than outcomes.** To encourage adaptation, countries should not unfairly favour certain interest groups, certain technologies, or certain forms of adaptation. By adopting institutions, they would enable people to make decisions about how to use resources – including their time, intelligence, labour and natural resources – without unnecessary bureaucracy. Encouraging political transparency and accountability, and eliminating corruption, would help – especially in countries where such aspects of government are currently lacking.

The institutions that support adaptation are fundamental to achieving our current and future priorities on global warming, as well as fostering resilience to any other problems that humanity may face in the future. 'Climate control' is thus short-sighted: while controlling the earth's climate may be motivated by the best of intentions, we may expose ourselves to other risks or we may simply neglect to deal with other risks of equal or greater concern.

We are best enabled to solve problems which might threaten humanity or our environment by prioritising our efforts, encouraging responsible decision making, focusing on the process rather than the outcome, and accounting for human error. Adaptation encourages us to have a view towards hypothetical problems in the future, but without neglecting problems that humanity faces today. Adaptation is thus the best way to ensure fairness and justice – for people today and for future generations.

Notes

1 **Could global warming bring mosquito-borne disease to Europe?**

1   'Global warming: Health and Disease' WWF Fact Sheet. Available at
    http://www.panda.org/resources/publications/climate/health_factsheet/
2   Levi-Strauss (1992), p. 425.
3   Wigley (1981); Lamb (1995); Chorley and Barry (1998).
4   Houghton *et al.* (1996); Tett (1999); Wigley and Schimel (2000).
5   Wigley (1981); Lamb (1995); Rampino (1987).
6   Lamb (1995), p. 433.
7   Calder (1974), p.143; Ponte (1976), p.306; Halacy (1978), p. 212.
8   Houghton *et al.* (1996); Tett (1999); Wigley and Schimel (2000).
9   Tett *et al.* (1999), pp. 569–72; Karl *et al.* (1996). pp. 279–92; Michaels and
    Knappenberger (1996), pp. 522–23; Kerr (1997), pp. 1040–42; Lindzen
    (1997), pp. 8335–42.
10 Ross (1991), p. 277; Gelbspan (1997), p. 278; Michaels and Balling (2000),
    p. 236.
11 See, for instance, McMichael *et al.* (1996); Patz *et al.* (1996); Hay (2001);
    Hay *et al.* (2002); Mouchet, J. *et al.* (1998); Epstein *et al.* (1998);
    McMichael, Patz and Kovats (1998); Martens (1998); Epstein (1999);
    Kovats *et al.* (1999); Reiter (2000); Shanks *et al.* (2000); Reiter (2001).
12 Gilles and Warrell (1993); Cook (ed.) (1996).
13 Bruce-Chwatt and de Zulueta (1980), p. 240.
14 Craig, *et al.* (1999), pp. 105–11.
15 Martens (1998), p. 176; Lindsay and Birley (1996), pp. 573–88; Jetten and
    Focks (1997), pp. 285–97; Patz (1998), pp. 147–53.
16 Rogers and Randolph (2000), pp. 1763–66; Dye and Reiter (2000), pp.
    1697–98.
17 Bruce-Chwatt and de Zulueta (1980), p. 240.
18 Lamb (1995), p. 433.
19 Homer (1990) Book 22, lines 31–37.
20 Lamb (1995), p. 433; Jones (1909).
21 Langholf (1990).
22 Bruce-Chwatt and de Zulueta (1980), p. 240.
23 Lamb (1995), p.433; Jones (1907).
24 Dante (1949) Cantica I: Hell (L'inferno). Canto XVII, lines 85–8; Chaucer
    (1977). 'The Nun's Priest's Tale', lines 134–140.
25 Campbell (1991), p. 232.

26 Bruce-Chwatt and de Zulueta (1980), p. 240.
27 Lamb (1995), p. 433.
28 Grove (1988).
29 Bruce-Chwatt and de Zulueta (1980), p. 240.
30 Harvey (1993), p. 91.
31 Bruce-Chwatt and de Zulueta (1980), p. 240.
32 *Quoted in* Bruce-Chwatt and de Zulueta (1980), p. 240.
33 Bruce-Chwatt and de Zulueta (1980), p.240; Dock (1927), pp. 241–47;
    Siegel and Poynter (1962), pp. 82–85; Dobson (1998), pp. 69–81; Poser and
    Bruyn (1999), p.165.
34 Defoe (1986).
35 Ekblom (1938), pp. 647–55.
36 Dobson (1980), pp. 357–89; Dobson (1989), pp. 3–7; Dobson (1994),
    pp. 35–60. Dobson (1997), p. 647.
37 Reiter (2000), pp. 1–11.
38 Lamb (1995), p. 433; Grove (1988).
39 Ekblom (1938), pp. 647–55.
40 Renkonen (1944), pp. 261–75.
41 Russell (1956), pp. 937–65.
42 Patz (1996), pp. 217–23; Epstein (1998), pp. 409–17.
43 Wesenberg-Lund (1921), pp. 383–86.
44 Ekblom (1938), pp. 647–55.
45 Bruce-Chwatt and de Zulueta (1980), p. 240; James (1920), pp. 71–85.
46 Bruce-Chwatt and de Zulueta (1980), p.240.
47 Crosnier (1953), pp. 1299–1388; Laigret (1953), pp.1308–12.
48 Galli-Valerio (1917), pp. 440–54.
49 Bruce-Chwatt and de Zulueta (1980), p. 240.
50 Hackett (1937), p. 336.
51 Bruce-Chwatt and de Zulueta (1980), p. 240.
52 Hackett (1937), p. 336.
53 Bruce-Chwatt and de Zulueta (1980), p. 240.
54 Brown, Haworth and Zahar (1976), pp. 1–25.
55 Gilles and Warrell (1993); Bruce-Chwatt (1987), pp. 75–110.
56 WHO (1978), pp. 9–17.
57 Lloyd and Coulter (1961).
58 Hackett (1937), p. 336.
59 Hackett and Missiroli (1935), pp. 45–109.
60 Bruce-Chwatt and de Zulueta (1980), p. 336.
61 de Jong, JCM (1952), pp. 206–9.
62 Bruce-Chwatt and de Zulueta (1980), p. 240.
63 Guido Sabatinelli, WHO, Regional Office for Europe, Copenhagen,
    Denmark.
64 Zucker (1996).
65 Sabatinelli (1998).
66 Watson, Zinyowera, and Moss (1996).
67 EPA (1997), pp. 1–4.
68 Manning and Nobre (2001), p. 74.

69   EPA (2002), pp. 1–4.
70   Epstein (2002).

2 Barriers to barriers: why environmental precaution has delayed mobile
   floodgates to protect Venice

1   James (2002).
2   WWF Italy 2003.
3   Norwich (1982).
4   Gentilomo andWarnock (1997), p. 2.
5   Day *et al.* (1999), p. 609.
6   Cecconi, Canestrelli, Corte and Di Donato (1997a), p. 1.
7   Gentilomo and Cecconi (1995), p. 435.
8   Cecconi (1997b), p. 5.
9   *Ibid.*, p. 16.
10   Ghetti (1988), p. 28.
11   Cecconi (1997c), p. 1.
12   Gentilomo and Warnock (1997), p. 8.
13   Day *et al.* (1996), p. 9.
14   Guthrie (2001).
15   Ghetti (1988), p. 26.
16   *Ibid.*, p. 27.
17   Zuccchetta (2000).
18   Cecconi (1997c), p. 3.
19   Cecconi (1997c), p. 2.
20   An historical tidal gauge on the island opposite St Mark's Square.
21   Fay and Knightley (1976).
22   'The rise in sea level caused by variations in atmospheric pressure and the
     wind occurs unexpectedly. This rise, calculated as the difference between the
     measured level and the astronomical tide, is known as a storm surge', in
     Cecconi (1997b), pp. 5–6.
23   Cecconi (1997c), pp. 4–5.
24   Agenzia Giornalistica Italia (2003).
25   International College of Experts (1998), p. 5.
26   Not all NGOs are against Project MOSE. Preservationist organisations
     based outside Italy, such as 'Venice in Peril' (UK) and 'Save Venice' (USA),
     have been generally positive about the mobile barriers. It is Italian
     environmental NGOs such as Italia Nostra, WWF Italy and Legambiente
     that have consistently opposed Project MOSE.
27   Brotto and Gentilomo (1998), p. 23.
28   International College of Experts (1998), p. 5.
29   *Ibid.* p. 46.
30   Gentilomo (1997), p. 32.
31   Zitelli and Rossetto (1996), p. 1.
32   Scarton, Perco and Borella (1996), p. 4.
33   Gentilomo (1997), pp. 35–36.

34 Gentilomo (1997), p. 34.
35 International College of Experts (1998), p. 31.
36 Gentilomo, M. *et al.* (1999), p. 14.
37 Quoted in Keahey, J. (2002), p. 240.
38 Communiqué of the Greens' Political Office, 28 April 2000 in Biorcio (2002), p. 55.
39 Anon (2003).
40 Baccaro (2003).
41 Fornasier (2003).
42 Vitucci (2003).
43 Standish (2003b).
44 Podger (2002).
45 Connor (2002).
46 Pirazzoli (1992).
47 Woodard (2003).
48 Ammerman *et al.* (1999).
49 Ammerman and McClennen (2000).
50 *Ibid.*
51 International College of Experts (1998), p. 17.
52 *Ibid.*
53 *Ibid.*
54 *Ibid.*
55 Connor (2002).
56 Brotto and Gentilomo (1998), p. 19.
57 Cecconi (1997c), p. 6.
58 Brotto and Gentilomo (1998), p. 24.
59 Cecconi (1997c), p. 6.
60 Brotto and Gentilomo (1998), p. 24.
61 Ammerman and McClennen (2000).
62 Italian Ministry of Public Works *et al.* (1997), p. 96.
63 International College of Experts (1998), p. 25.
64 Ammerman and McClennen (2000).
65 *Ibid.*
66 Cecconi (1997c), p. 5.
67 *Ibid.*, p. 6.
68 Runca *et al.* (1996), p. 16.
69 International College of Experts (1998), p. 6.
70 *Ibid.*, p. 6.
71 Cecconi (1997c), p. 4.
72 Gentilomo (1997), p. 5.
73 Standish (2001b).
74 Goklany (2001), p. 2.
75 Zamparutti (2003).
76 I have explored elsewhere how Italian social policy for environmental management and beyond has become increasingly governed by the precautionary principle. See Standish (2001a).
77 Gentilomo *et al.* (1999), p. 11.

78   Mazzacurati (1996), pp. 1–2.
79   Berstein (1998), pp. 48–50.
80   Standish (2002b).
81   Italia Nostra (no date).
82   Standish (2003a).
83   Standish (2002a).
84   Morris (1993), p. 11.

3 Climate change: the 21st century's most urgent environmental problem or
   proverbial last straw?

1   Clinton (1998). *See also*, Woodwell (1997), Greenwire (1998).
2   Trenberth (undated), Knowlton (2000). In fact, the Dutch government
    commissioned an IMAX movie by IRAS Films, *The Straw that Breaks the
    Camel's Back*, for screening at the World Climate Conference in The Hague,
    November 2000. See http://www/irasfilm.com/200101straw.htm
3   IPCC (2001a), pp. 2–3.
4   Goklany (2001a); Goklany (2002c).
5   FAOSTAT Database.
6   The 1969–71 estimate is from FAO (1996). The 1998–2000 figures are
    from FAO (2002): 31.
7   World Bank (2002).
8   Goklany, (2001a); Goklany, (2002c).
9   Goklany (2000a); Goklany (1999b).
10 IPCC (2001a), p. 31.
11 *Ibid.*
12 McNeely, *et al.* (1995), pp. 755–57.
13 Goklany (2002b).
14 Goklany (1998); Goklany (2000).
15 FAO (2000).
16 Walther *et al.* (2002).
17 Parmesan and Yohe (2003). *See also*, Root *et al.* (2003).
18 Parmesan and Yohe.
19 WWF Finland (2002).
20 RSPB (2000), p. 14; RSPB (2001), pp. 19–20; *Wildlife News* (2002).
21 Fox (2001a); Fox (2001b).
22 *Ibid.*
23 Fitter and Fitter (2002).
24 Walther *et al.* (2002), p. 392; Vaughan *et al.* (2001).
25 Walther *et al.* (2002).
26 Goklany (2001a).
27 Myneni *et al.* (1997); N. Nicholls (1997).
28 Ausubel (1991).
29 Goklany (2000); Goklany (1995); Goklany (1992).
30 Arnell *et al.* (2002), p.418.
31 IPCC (2001a): p. 13.

32 Hulme *et al.* (1999), p. S8, S14.
33 Wigley (1998); Malakoff (1997).
34 Goklany (2003).
35 Parry *et al.* (1999), pp. S60, S62. *See also*, Gitay *et al.* (2001), p. 259.
36 Goklany (1998).
37 Goklany (1999a).
38 Arnell *et al.* (2002), p. 439.
39 *Ibid.*; McMichael *et al.* (2001), p. 466.
40 In 1990, about a tenth of the population estimated to be at risk contracted
    malaria, while fatalities were about 0.2% of that. *See* WHO (1999).
41 Arnell (1999), Table 5.
42 *Ibid.*: Table 6.
43 Arnell *et al.* (2002), p. 424.
44 Solomon *et al.* (1996), pp. 492–96.
45 Hulme *et al.* (1999), p.S14.
46 IPCC (2001a), p. 73; Henderson-Sellers, *et al.* (1998a); Henderson-Sellers,
    (1998b).
47 This section draws heavily upon: Goklany (2001b), pp. 465–74; and
    Goklany (2000).
48 WHO (1999), p. 56.
49 Goklany (2002a).
50 Goklany (2000); Goklany (2001a).
51 Goklany (1995); Goklany (2000).
52 FAOSTAT Database.
53 Goklany (1995); Goklany (1999b); Goklany (2000).
54 IPCC (2001b), p.10.
55 World Bank (2002).
56 IPCC (2001b), pp. 238, 240, 259, 295.
57 *Ibid.*, pp. 957–58.
58 Pearce *et al.* (1996), p. 191.
59 Goklany (2000).
60 Byerlee and Echeverria (2002).
61 Goklany, (1998); Goklany (1999a).
62 Ha-Duong *et al.* (1997); Wigley *et al.* (1996); Wigley (1997).
63 Goklany (2000).
64 If they are not costly in a socioeconomic sense, they are, almost by
    definition, no-regret actions.

4 Is Kyoto a good idea?

1 http://www.ipcc.ch/about/about.htm
2 http://www.grida.no/climate/ipcc_tar/index.htm
3 IPCC (2001a), p. 5.
4 IPCC (2001d), chapter 1.3.1.
5 Ibid, Chapter 7, executive summary.
6 IPCC (2001c), section D.1.

7   *Ibid.*, Chapter 7, executive summary.
8   Lindzen *et al.* (2001).
9   Nature News Service, 7 March 2001, http://www.nature.com/nsu/010308/010308-9.html
10   IPCC (2001b), p. 1.
11   IPCC 2001, Working Group 1, Chapter 7, Executive Summary.
12   IPCC (2000).
13   Henderson (2003).
14   Corcoran (2002). See Webster *et al.* (2001) for a more in-depth analysis of projected emissions of carbon dioxide.
15   *Ibid.*
16   Castles and Henderson (2003).
17   IPCC 2000, Summary for Policymakers, p. 3.
18   NAS (2001).
19   Lomborg (2001), p. 290, citing IPCC (1998), p.7.
20   Titus *et al.* (1991).
21   IPCC (2001e).
22   Lomborg (2001), p.293, citing Bove (1998); Pielke and Landsea (1999).
23   IPCC (2001b), p. 15.
24   See Chagnon *et al.* (2000) and Kunkel *et al.* (1999).
25   See Gouk *et al.* (1999), Fernandez *et al.* (1998).
26   Goklany (2000), p. 198.
27   Lomborg (2001), p. 288, Table 7, citing IPCC (1996), p. 451, Rosenzweig and Parry (1994), p.136.
28   'Night minimum temperatures are continuing to increase, lengthening the freeze-free season in many mid- and high-latitude regions.' IPCC (2001), Executive Summary, Chapter 2, 'Observed Climatic Variability and Change', available at http://www.grida.no/climate/ipcc_tar/wg1/049.htm
29   Keatinge *et al.* (2000).
30   Goklany (2000), p. 197.
31   *Ibid.*, p. 196.
32   See http://unfccc.int/cop7/
33   See http://www.ieta.org, especially: http://www.ieta.org/Documents/New_Documents/StateandTrendsoftheCarbonMarket2002.pdf
34   Nordhaus (2001), Global Warming Economics, *Science*, vol. 294, 9 November 2001.
35   Grubb (2003).
36   Nordhaus (2001), Global Warming Economics, *Science*, vol. 294, 9 November 2001.
37   McKibbin and Wilcoxen (2003).
38   Institut for Miljøvurdering, 'Danmarks omkostninger ved reduktion af CO_2' (October 2002). Link: www.imv.dk
39   The Liberal and Conservative Party form a coalition minority government backed by the nationalist Danish Peoples Party.
40   Weyant and Hill (1999).
41   IPCC (2001g), Chapter 8.2.2.1.1, table 8.4.
42   Nordhaus (2001).

43   Institut for Miljøvurdering, 'Danmarks omkostninger ved reduktion af CO_2' (October 2002), p. 23. My translation. Link: imv.dk
44   Nordhaus and Boyer (1999).
45   *Ibid.*, p. 33 – The model run assumes US participation. This is a reasonable assumption since Kyoto could not survive US non-participation in the long run.
46   McKibbin and Wilcoxen (2003).
47   *Ibid.*
48   Nordhaus and Boyer (1999).
49   *Ibid.*, question 6.3, p. 19.

5 Sustainable energy for the poor

1   See http://www.choose-positive-energy.org/html/content/facts_backgrnd.html
2   Sobhani and Retallack (2001), p. 225.
3   IEA (2002), p.14.
4   *Ibid.*, p.16
5   Goldemberg and Johansson (1995).
6   EIA (2001).
7   Smith (2000).
8   Ravindranath (2000), p. 30.
9   See 'Rural Energy in India' at http://www.incg.org.in/CountryGateway/RuralEnergy/Overview/RuralenergyinIndia.htm
10   UNDP (2000), p. 8.
11   Ravindranath and Hall (1999), p.19.
12   Smith (2000).
13   WHO (2000).
14   Bruce *et al.* (2002).
15   Smith and Mehta (2000).
16   Smith (2000).
17   This estimate uses a disability-adjusted lost life-year approach. *Ibid.*
18   Bruce, *et al.* (2002).
19   WHO (no date).
20   Narain (2003).
21   Bruce *et al.* (2002).
22   *Ibid.*
23   WHO (no date).
24   Ravindranath and Hall (1999), p. 44.
25   *Ibid.*
26   *Ibid.*, p. 52.
27   Auckland *et al.* (2002), p. 2.
28   Subudh (1993), p. 107.
29   Ravindranath and Hall (1999), p. 76.
30   Sukla (1997), p. 368.
31   TERI (2002).

32   Guru (2002.
34   DFID (2002), p. 12.
35   World Bank (2002), Sections 3.6–3.8.
36   DFID (2002), p. 18.
37   Kyoto Protocol, Article 12, Paragraph 2.
39   See 'Free-Riders and the Clean Development Mechanism', available at http://www.panda.org/downloads/climate_change/freeriders.rtf
40   Climate change and India, p. 222.
41   DFID (2002), p. 12.

6 Energy for the poor? The clean development mechanism

1   Nelson (2003).
2   This subject is explored in depth by the late economist Simon (1981) and Lomborg (2001).
3   Calder (1975).
4   Lemonick (1994).
5   Isotope analysis shows that the extra carbon has no C-14 and therefore must come from hydrocarbon fuels. The 420,000-year record comes from the Vostok ice core.
6   For example, Keigwin (1996). The increase in temperatures in northern Europe caused a large increase in agricultural production, triggering various historical movements, including the Viking invasions.
7   ERI (2002).
8   Von Schirnding *et al.* (1991).
9   Dr Philip Lloyd, Energy Research Institute, University of Cape Town.
10   Paraffin Safety Association of South Africa (2001).
11   Studies by Dr Philip Lloyd, Energy Research Institute, University of Cape Town. From the National Electricity Regulator.
12   *Ibid.*
13   'Severe Accidents in the Energy Sector'. Paul Scherrer Institut. PSI Bericht Nr 98-.
14   Nuclear reactions make Plutonium 239, which is a fissile material that can be used in bombs. But soon afterwards, they make Plutonium 240, which is not fissile and which contaminates the 239, making it useless for bombs. So the waste fuel must be extracted quickly to make bombs. Most power reactors run for a year or so before re-fuelling, making their waste useless for bombs.
15   'Comparison of energy sources in terms of their full energy chain emission factors of greenhouse gases', Joop F. van de Vate. *Energy Policy*, vol. 25, No 1. Elsevier.
16   In 1995, South Africa emitted 310 million tons of carbon dioxide equivalent, of which 140 million tons came from coal power stations. ERI.
17   UNDP (1998).

7 Warming aid, chilling trade?

1   Aaron Wildavsky was Professor of Political Science at the University of California, Berkeley, and served as the President of the American Political Science Association. Professor Wildavsky died in 1996. Wildavsky (1992).

2   At the time of writing, but this may change subsequently.

3   The phrase cunning plan is now closely associated in (British) English with the character Baldrick from the television series *Black Adder*. The association is not an unhappy one.

4   Even with the Vienna Convention, however, it turned out to be very difficult to establish a system that was equitable and the political process was gamed by vested interests in the form of DuPont, which lobbied for a more rapid phase-out than might have been optimal because it had developed patented alternatives to the CFCs. Also, India, China and Russia soon started producing large quantities of CFCs, not only for their home markets but also for illegal sale to overseas markets.

5   Barrett (1994).

6   WHO (2002).

7   'So it's worth remembering (however many revisionist interpretations of it there may now be!) that there was undoubtedly a "deal" on the table at the Earth Summit. G77 and emerging countries implicitly agreed to sign up to a variety of action plans for addressing some of the big environmental issues (global warming, deforestation, loss of biodiversity etc.), whilst OECD countries implicitly signed up to the idea of increased aid flows and other forms of development assistance as the quid pro quo for their buy-in on the environment agenda.' ECO (1997).

8   Indeed the agreement might more accurately be titled the convention on cash for despots but for the fact that very little money has flowed as a result. (*See* e.g. Morris 1995).

9   This amounts to barely one gold-plated Rolls-Royce car per corrupt government official. (The calculation is a simple one: assume that a gold-plated Rolls-Royce costs about $200,000; so $2 billion will buy 20,000. Assume further that the money is disbursed to 100 countries; that means 200 officials in each country get a gold-plated Rolls-Royce.)

10   *See* Morris (2000).

11   't Sas-Rolfes (1995).

12   The following discussion is adapted from Morris (2000), pp. 267–301.

13   That is to say, when the cost of producing all goods in country A is greater than in B, there will still be trade if the *relative* cost of producing some goods is greater in B than in A. Thus, in Ricardo's famous example, although both wine and cloth may be more expensive to manufacture in Britain than in Portugal, the relative cost of producing wine compared to cloth is greater in Britain than in Portugal, so people in Britain will sell cloth to Portugal in exchange for wine.

14   Pearce (1994), pp. 20–38.

15   The report was never adopted by the GATT contracting parties, so the decision is only of value as guidance. Appleton (1999), p.206.

16   *United States – Measures Affecting Alcoholic and Malt Beverages* (1992).

Report of the panel, GATT document DS23/R, adopted 19 June 1992.

17   Even if extra jurisdictional application were permitted, the DP said that the measures in question were not necessary within the meaning of Article XX (b) because the US had not demonstrated that it had exhausted all reasonable GATT-consistent options, such as the negotiation of an MEA to protect dolphins.

With regard to XX(g), the DP noted that since a country can only control the production or consumption of a natural resource if that production or consumption is under its jurisdiction, Article XX(g) should not be applied extrajurisdictionally.

Even if they could be applied extrajurisdictionally, the panel decided that the US measures did not meet the requirement of Article XX(g) that they 'relat[e] to the conservation of exhaustible natural resources'. To *relate to* conservation, the panel said, the measures must be 'primarily aimed at' such conservation. Because of the unpredictability caused by the linkage between the permitted Mexican incidental taking rate and the actual US taking rate, the US measures could not be considered 'primarily aimed at' the conservation of dolphins.

Applying the same reasoning, the panel decided the *intermediary* nation embargo could not be justified under Article XX(b) or (g). See Goldberg (1994).

18   Appleton, *supra* note 4, p. 206.

19   Hogue (1996) [1985].

20   DP report, cited at http://www.american.edu/projects/mandala/TED/ TUNA2.HTM

21   For a commentary see e.g. http://www.american.edu/projects/mandala/TED/ TUNA2.HTM

22   The myth of smart dolphins is ancient and probably rests on a mistaken view that dolphins care for humans: it is found in Greek mythology, in which a dolphin saves a man from drowning; there are numerous reports of such incidents, which are most probably the result of a dolphin mistaking the man for a drowning dolphin.

23   See http://www.defenders.org/wildlife/dolphin/tundolph.html; http://www.nwf.org/trade/dolphins.html

24   This case particularly irritated the environmentalists, in part because the US had attached three *amicus curiae* (friend of the court) briefs submitted by environmentalist NGOs, which the panel chose not to consider. The panel's grounds for so doing were that it had not requested the briefs and that they did not form an integral part of the requested submission of the US government.

25   The AB also overruled the panel on the point regarding *amicus* briefs, arguing that under certain circumstances dispute panels would be obliged to accept such briefs.

26   Section 609 was applied without prior negotiation with the four Asian countries, whereas with other countries it had been applied with prior negotiation; also, the timeframe for implementation was shorter for the Asian countries. Finally, the AB noted that the mode of certification was

lacking in transparency and predictability, thereby denying all nations subject to the regulation basic fairness and due process.

27 *United States – Import Prohibition of Certain Shrimp Products (1998)*, para. 185, p. 75.

28 Cuba currently catches approximately 5,000 sea turtles per year, a perfectly sustainable level; yet it is facing serious opposition from environmentalists to a proposal to trade in 500 of these – a trade that would, by increasing the value of turtles, make it more worthwhile investing in such things as turtle hatcheries.

29 Robert Nappier of the WWF was quoted as supporting such a proposal at a meeting at Chatham House in London (http://www.re-focus.net/jun2001_3main.html accessed 17 July 2003).

30 *See* http://www.uscib.org/%5Cindex.asp?documentID=2496

31 Specifically, UNICE has proposed that as a rebuttable presumption, the WTO accept in principle the validity of trade measures contained in MEAs (presumably trade measures that are discriminatory in nature and those that are consistent with WTO rules). This rebuttable presumption essentially reverses the burden of proof and raises the standard of proof in respect of compatibility between a trade measure pursuant to an MEA. In other words, any party to the GATT who believes that a trade measure is being pursued in a manner that is in violation of the *chapeau* of Article XX should have to show positively that the measure is in fact in violation of its rights and should be required to provide substantial evidence of this fact. And that 'MEAs reflect a broad consensus in the international community on how to solve global environmental issues.'

32 Doha Declaration, available at: http://www.wto.org/english/thewto_e/minist_e/minol_e/minded_e.htm

33 This chapter perhaps also offers a framework for assessing the validity of such claims.

8 How Europe's risk regulations affect business

1 European Commission (1995).

2 European Commission (1997).

3 Lomborg (2001), pp 258–323.

4 Lomborg (2001). p. 302. Lomborg refers to three separate models: Parry *et al.* 1998: 286, Nordhaus and Boyer 1999: 104 and WEC (1998).

5 DRI–WEFA 2002; ICCF (2002).

6 European Commission (2003).

7 EEA (2002); EEA (2003).

8 Macalister (2003).

9 ICCF (2002).

10 Shell (2002), p.28.

11 GCEP website, http://gcep.stanford.edu/project_detail/fact_sheet.html

12 'Call for Action', http://archive.greenpeace.org/earthsummit/wbcsd/

13 'Green or Greenwash? A Greenpeace Detection Kit'

http://archive.greenpeace.org/~comms/97/summit/greenwash.html
14   'Challenging Corporations: Introduction to FOE's Corporates Campaign'
     http://www.foe.co.uk/resource/briefings/corp_alert_intro_lgs.pdf
15   FOEI (2001).
16   Desrochers (2002), pp.53–54.
17   *Ibid.*

9 Bootleggers, Baptists and the global warming battle

1   COP6 proceedings can be found at the Conference website:
     http://cop6.unfccc.int/ (last visited 12 July 2001).
2   The rejection originated in a letter from President Bush to several senators.
     See White House (2001).
3   The US Senate passed a resolution on July 25 1997, by a margin of 95-0,
     requesting the executive branch not to sign the Kyoto Protocol unless a
     commitment was made by developing countries to reduce emissions and if
     the Protocol was shown not to cause serious harm to the US economy. US
     Senate (1997).
4   *The Economist* (2001a).
5   Steyn (2001).
6   Browne (2001).
7   *Ibid.*
8   Easterbrook (2001).
9   *Ibid.* Easterbrook adds that 'American commentators have happily parroted
     Europe's line.'
10   *See* EIA (1998). For a critique of this study, see Geller (1998).
11   This theory was first described in Yandle (1983); *see also* Yandle (1999).
12   Kyoto Protocol Art. 3, § 1.
13   *See* US DOE (no date).
14   Kyoto Protocol. Art 3, § 2.
15   *Ibid.* Art. 3, § 5.
16   *Ibid.* The Conference of the Parties is the Protocol's term for the 'supreme
     body of the Convention'. *Ibid.* Art. 13 § 1.
17   *Ibid.* Art. 3, §§ 10, 11.
18   *Ibid.* Art. 6, § 1(b).
19   *Ibid.* § 1(c).
20   *Ibid.* § 1(d).
21   *See* Black-Arbeláez (2001).
22   Kyoto Protocol. Art. 4, § 1.
23   *See* EU (2000).
24   The International Climate Change Project Fund, sponsored by the United
     States Agency for International Development, is currently seeking to fund
     joint implementation projects in USAID-assisted countries in Asia, Africa
     and Latin America. For more information, see Joint Implementation Online,
     http://www.ji.org
25   Kyoto Protocol, Art. 6, § 1.

26 *Ibid.* Art. 12.
27 *Ibid.* Art. 12, § 5(b), (c). For further discussion of the Clean Development Mechanism, see Hourcade and Toman (1999).
28 The exclusion of China alone may undermine any possibility of keeping worldwide emissions from growing. *See* Deborah E. Cooper, Note, 'The Kyoto Protocol and China: Global Warming's Sleeping Giant', 11 *Georgetown International Environmental Law Review* 401 (1999).
29 *See* Manne and Richels (1999), pp. 1, 20; Reid & Goldemberg (1997), p. 233.
30 NEIC (2001).
31 Malakoff (1997).
32 BBC News (2000).
33 *See, e.g.*, Bolin (1998), pp. 330–31. Another scientist (Wigley 1998) estimated that if all nations met their Kyoto obligations, the likely reduction in global warming by year 2050 would be even less: 0.07°C.
34 *See, e.g.*, Bradsher and Revkin (2001).
35 Gordon Reid Smith of BP has said, 'Every reduction in energy use directly translates to reductions in operating costs in products available for sale along with decreases in combustion CO_2 emitted.' Shook (1999).
36 *See generally* Samuelson (1954).
37 *See* Yandle (1983).
38 Indeed, Kyoto enjoys the support of a variety of religious groups, including some actual Baptists. For example, the South Carolina Interfaith Climate Change Campaign includes Catholics, Methodists, Lutherans, Presbyterians, and the Cooperative Baptist Fellowship; Munday (2000). Joan Brown Campbell, head of the National Council of Churches, has said that she wants to make support for Kyoto a 'litmus test for the faith community'; Cushman (1998).
39 Nelson (1993), pp. 233, 234.
40 Stott (2001). Stott is emeritus professor of biogeography at the University of London.
41 Breyer (1982), pp. 269–70.
42 Along the same lines, Leidy and Hoekman (1996) suggest that environmental activists prefer standards over taxes because the 'flexibility and autonomy remaining in the hands of polluting firms under a penalty tax is undesirable'; pp. 43, 54.
43 As Frances Cairncross (1995) notes, 'When the richer countries have offered to meet their goals for carbon dioxide cuts partly by paying for energy-saving measures in the developing world, the poorer countries have sometimes accused them of trying to buy their way out of their environmental responsibilities.' p. 73.
44 The exploitation, however, might not be all one-way – if the number of carbon permit sellers is few enough (that is, if only a small number of countries have emissions low enough that they can sell permits to other countries), there would arise a 'considerable potential for extracting monopoly rents'; Manne and Richards (1998). If, as seems likely, the former Soviet states (especially Russia and the Ukraine) dominated the market for

permits, the 'expected efficiency gains from establishing a permits market among Annex I countries could be reduced by about a third'; (Burniax 1998).

45   Raven (1998).
46   Cordato (1999).
47   *See* Black-Arbeláez (2001), p. 35, Table 12.
48   *Ibid.*
49   Jorgenson and Wilcoxen (1993).
50   *Ibid.*
51   Manne and Richels, *supra* note 141.
52   DRI–McGraw-Hill (1997).
53   *Ibid.*
54   *Ibid.*
55   *Ibid.*
56   *Ibid.*
57   *Ibid.*
58   *Ibid.*
59   WEFA, Inc. (1998), p. 2.
60   *Ibid.* at 4–5.
61   Manne and Richels (1998).
62   WEFA, Inc. (1998), p. 4.
63   Wallsten (2000), p. 12.
64   *Ibid.*
65   Also the World Bank reportedly invested nearly $16 billion since 1992 in oil, gas, coal and other power projects around the world, particularly in Third World countries. See http://www.seen.org/pages/press_releases/pr_leak.shtml
66   See Wysham *et al.* (1999), p. 5.
67   *Ibid.*
68   Energy Report (2000).
69   See Exxon-Mobil (2000), Exxon-Mobil (2001).
70   Hammitt (2000). Hammitt is an associate professor of economics at the Harvard Center for Risk Analysis.
71   Lavelle (2001).
72   Frances Cairncross observes that 'Britain's nuclear electricity generators have been keen on carbon tax: not surprisingly, as they are the main commercial source of carbon-free energy.' Cairncross (1995), p. 192.
73   Kakuchi (1997).
74   Reuters (2001b).
75   Reuters (1999).
76   Reuters (2001a).
77   John (2001).
78   Reuters (2001b).
79   *See, e.g.*, Reuters (2000).
80   Utility Environment Report (1998).
81   *The Economist* (1997).
82   For a chart showing the difference between Kyoto commitments and actual

1995 emission levels for various countries, see Grubb, Vrolijk & Brack (1999), p. 162.

83   *Oil and Gas Journal* (2000).
84   ENS (2000).
85   *See also The Economist* (1997).
86   Revkin (2001).
87   *See, e.g.*, McKibbin *et al.* (1999), p. 287. Environmental activists, of course, disagree. *See, e.g.*, Repetto & Maurer (1997).
88   Cordato (1999).
89   Castle (2001).
90   See Laird (2000).
91   *See, e.g.*, Drozdiak (2000).
92   Horner (2000).
93   Fan *et al.* (1998).
94   *See, e.g.*, Grubb *et al.* (1999), pp. 79–80.
95   *See* Topping Cone (2000). ('The EU, with significant limits on land available for reforestation by its member countries, is lobbying for these credits to have a lower value in an emissions trading system ... ')
96   *The Economist* (2001b).
97   Loy (2000).
98   *See generally* FT Energy Newsletters (2001).
99   *See* Victor (2001).
100 LeVine (2000).
101 *Ibid.*
102 *See* Kyoto Protocol, Art. 4, § 1; *see also supra* notes 39–40 and accompanying text.
103 *See ibid.* at 85.
104 Witter (1997).
105 *Ibid.* (statement of Melinda Kimble, Acting US Assistant Secretary of State for the Environment).
106 *Ibid.*; *see also* Utility Environment Report (1997).
107 Yandle (1999), p. 35.
108 Planet Ark (2000).
109 *Ibid.*
110 *See, e.g.*, Brown (2000).
111 For a good description of this difficulty, see Mitchell & Chayes (1995), pp. 115, 120–27.
112 Laird (2000).
114 Cooper (1998), p. 66.
115 *Ibid.*
116 *Ibid.*

10 Climate change and civilisation collapse

1   Weiss and Bradley (2001).
2   Weiss and Bradley (2001).

3   DeMenocal (2001); Mapes (2001).
4   Claussen *et al.* (1999).
5   Tainter (1990).
6   Pointing (1991).
7   Keys (1999).
8   Baillie (1999).
9   Berglund (2003).
10   Weiss (2000).
11   Weiss (2001).
12   Weiss *et al.* (1996); Weiss (2000).
13   Peiser (1998).
14   See http://research.yale.edu/leilan/ for more information about this project.
15   Courty (1998).
16   Butzer (1996).
17   Baillie (1998).
18   Butzer (1996) and Baillie (1998).
19   Baillie (1998).
20   Manning (1996).
21   Butzer (1996); Possehl (1996).
22   Weiss (2000).
23   Gill (2000).
24   *Ibid.*
25   Webster (2000).
26   *Ibid.*
27   Haug *et al.* (2003).
28   Webster (2000).
29   Weiss (2001).
30   Engvild (2003).

11 The political economy of climate change

1   See, for example, the 'Galileo 2001 – For the freedom and dignity of science' manifesto, http://www.cidis.it/articoli/vari/galileo2001.htm; Gaspari (1997), pp. 234–35; Gheddo and Beretta, p. 227.
2   See Ricci (2002), p. 39.
3   Francescato (2002), p. 43.
4   IEA (2001). See also EIA (2002).
5   Brunetti, *et al.* (2000); Moonen *et al.* (2002).
6   Tognetti *et al.* (1998).
7   See Bianco *et al.* 2002.
8   See Midena (2002), p.7; Serafini (1998).
9   EIRO (no date).
10   Lomborg (2001), p. 260, citing IPCC (2001) Section 3.1.
11   Marsh (2002).
12   Baker (2001), p.138.
13   Jones *et al.* (2001).

14   Ortolani (2000), p.5.
15   This theme was explored recently in an analysis of proxy data from the past
      1,000 years, which concluded that 'Many records reveal that the twentieth
      century is likely not the warmest nor a uniquely extreme climatic period of
      the last millennium.' Soon *et al.* (2003), p. 233.
16   Friis-Christensen and Lassen (1991).
17   Corbyn and Golipur (1996).
18   McKitrick (2001).
19   Spencer and Christy (1990), p. 1558.
20   Battaglia (2000).
21   'Today's 30 percent increase in atmospheric concentrations of CO_2 is
      estimated to have increased crop output between 5 and 10 percent.
      Doubling CO_2 concentrations could increase growth of the same plants and
      crops by as much as 30 percent', Bradley (2000), p. 92. 'Carbon dioxide is
      not a pollutant; it is essential to life. Based on extensive evidence from
      agricultural research on enhanced carbon dioxide environments both in the
      field and in laboratories, increases in carbon dioxide should cause many
      plants to grow more vigorously and quickly. The reason is that most plants
      evolved under, and so are better adapted to, concentrations of atmospheric
      carbon dioxide higher than those found at present', Soon and Baliunas, *et
      al.* (2001), p. 37.
22   Baker (2001), pp. 146–47. Emphasis added.
23   Castles (2002).
24   Crandall (1997), p. 145.
25   Lomborg (2001), p. 301, citing IMF (2000), p.113 ('with global GDP at
      32,110 billion dollars in 2000').
26   FAO (1991).
27   Simon (1996).
28   Cohen (1995), pp. 576–87.
29   Goklany (1995), p. 442.
30   Hardin (1968); Mitchell and Simmons (1994); Steele (2002).
31   'The goal of the Protocol is to stabilise *emissions* of CO_2, not the
      atmospheric *concentrations* of CO_2 (and of course the other greenhouse
      gases). Even if emissions could be stabilised at 1990 levels, six billion tons
      of carbon would be added to the atmosphere annually by human activities.
      That carbon would build up in the atmosphere and a doubling of CO_2
      would still occur near the middle of this century.' Balling (2002), p. 156.
32   Kyoto Protocol (1997).
33   Thorning (1998).
34   Goklany (2001a).
35   Goklany (2001b), p. 73.
36   Lomborg (2001), p. 302, citing Parry *et al.* (1998), p.286, WEC 1998,
      Nordhaus and Boyer (1999), p.104.
37   Montgomery (1997), pp. 65–68.
38   DRI–WEFA (2002).
39   *Ibid.* p. 15.
40   EEA (2003).

41   Thorning (2002).
42   Buchanan and Tullock (1975), pp.139–47.
43   See Reisman (2002), pp. 13–14, for a discussion of how Austrian economists might address global warming. Desrochers (2002).
44   Meiners and Yandle (1998), pp.63–64. See also Rothbard (2002), pp. 260–68; Leoni (1991).
45   Cordato (1999).
46   See Cordato (1999), pp. 5, 10.
47   *Ibid.* p.4. See also Buchanan (1969).
48   Desrochers (2002).
49   See Simon (1996), pp. 54–72; Bradley (2002).
50   Yandle (1999).

Epilogue

1   Pearce (1991), p. 284.
2   *Ibid.*
3   Pearce (1991), p. 287.
4   FOE UK (2001), p. 9.
5   Retallack and Sobhani (2002), p. 65.
6   IFG (2002), Alternatives to Economic Globalization, Spring.
7   Dresner (2002), p. 58.
8   FOE UK (2001), p. 17.
9   Louw (2002).
10   FOE UK (2001).
11   De Soto (1989), p. 243.
12   *Daily News*, (Sri Lanka) (2003), 'Preparing for a future deluge', 2 July; http://www.dailynews.lk/2003/07/02/fea01.html
13   European Commission (2002).

Bibliography

1 Could global warming bring mosquito-borne disease to Europe?

Brown, A. W., Haworth, A. J. and Zahar, A. R. (1976), 'Malaria eradication and control from a global standpoint', *Journal of Medical Entomology*, vol. 13, pp. 1–25.

Bruce-Chwatt, L. J. (1987), 'Malaria and its control: present situation and future prospects', *Annual Review of Public Health*, vol. 8, pp. 75–110.

Bruce-Chwatt, L. J., and de Zulueta, J., *The Rise and Fall of Malaria in Europe, a Historico-epidemiological Study* (Oxford, Oxford University Press, 1980).

Campbell, B. M. S. (ed.), *Before the Black Death: Studies in the 'Crisis' of the Early Fourteenth Century*, (Manchester, Manchester University Press, 1991).

Calder, N., *The Weather Machine* (New York, NY, Viking, 1974).

Chaucer, G., *The Canterbury Tales; translated into Modern English by Nevill Coghill*, (London, Penguin, 1977).

Chorley, R. J., and Barry, R. G., *Atmosphere, Weather and Climate* (New York, NY, Routledge, 1998).

Cook, G. (ed.), *Manson's Tropical Diseases* (London, W. B. Saunders Co., 20th edn, 1996).

Craig, M. H., Snow, R. W. and le Sueur, D. (1999), 'A climate-based distribution model of malaria transmission in sub-Saharan Africa', *Parasitology Today*, vol. 15, no. 3, pp. 105–11.

Crosnier, F. (1953), 'De quelques considérations sur le paludisme métropolitain', *Revue Path. hum. comp.* 53, pp. 1299–1388.

Dante, A., *The Comedy of Dante Alighieri the Florentine*, (London, Penguin [reprinted]).

de Jong, J. C. M., 'The influence of changes in chlorine content of inland waters on malaria in Friesland', *Documenta Med. Geogr. Trop.*, vol. 4 (1952).

Dobson, M. J. (1980), 'Marsh Fever' – the geography of malaria in England', *Journal of Historical Geography*, vol. 6, pp. 357–89.

Dobson, M. J. (1989), 'History of malaria in England', *Journal of the Royal Society of Medicine*, vol. 82 (Suppl 17), pp. 3–7.

Dobson, M. J. (1994), 'Malaria in England: a geographical and historical perspective', *Parassitologia*, vol. 36, nos. 1&2, pp. 35–60.

Dobson, M. J., *Contours of death and disease in early modern England* (Cambridge, Cambridge University Press, 1997).

Dobson, M. J. (1998), 'Bitter-sweet solutions for malaria: exploring natural remedies from the past', *Parassitologia*, vol. 40, nos. 1&2, pp. 69–81.

Dock, G. (1927), 'Robert Talbor, Madame de Sévigné, and the introduction of cinchona. An episode illustrating the influence of women in medicine', *Annals of Medical History*, vol. 4, pp. 241–7.

Dye, C. and Reiter, P. (2000), 'Temperatures without fevers?' *Science*, vol. 289, pp. 1697–8.

Dymowska, Z., '*Zimnica* ['Malaria']', in *Choroby zakazne w Polsce i ich zwalczanie w latach 1919–1962* [Infectious diseases in Poland and their control, 1919–1962], J. Kostrzewski (ed.) (Warsaw, Panstwowy Zaklad Wydawnictw Lekarskich, 1964).

Ekblom, T. (1938), 'Les races Suédoises de l'Anopheles maculipennis et leur role épidémiologique', *Bull. Soc. Pathol. Exot.*, vol. 31, pp. 647–55.

Epstein, P. R., *et al.* (1998), 'Biological and physical signs of climate change: focus on mosquito-borne diseases', *Bulletin of the American Meteorological Society*, vol. 79, pp. 409–17.

Epstein, P. R. (1999), 'Climate and Health', *Science*, vol. 285, pp. 347–48.

Epstein, P.R., *Global Warming: Health and Disease* (WWF, 2002).

Environmental Protection Agency (EPA), *Climate Change and Public Health* (Office of Policy, Planning and Evaluation 2171, Washington DC, EPA, 1997).

EPA, *Global Warming – Impacts. Health* (Office of Policy, Planning and Evaluation, Washington, DC EPA, 2002).

Galli-Valerio, B. (1917), *La distribution géographique des Anophèlines en Suisse au point du vue du danger de formation de foyers de malaria*, Bull. schweiz. Gesundheit Amt., vol. 40, pp. 440–54.

Gelbspan, R., *The heat is on: the high stakes battle over Earth's threatened climate* (Reading, Massachusetts, Addison–Wesley, 1997).

Gilles, H. M. and Warrell, D. A., (eds), *Bruce-Chwatt's Essential Malariology* (London, Edward Arnold, 1993).

Grove, J. M., *Little Ice Age* (London, Routledge, Keegan and Paul, 1988).

Hackett, L. W. and Missiroli, A. (1935), 'The varieties of *Anopheles maculipennis* and their relation to the distribution of malaria in Europe', *Riv. Malar*, vol. 14, pp. 45–109.

Hackett, L. W. *Malaria in Europe, an Ecological Study* (London, Oxford University Press, 1937).

Halacy, D. S., *Ice or Fire? Can We Survive Climate Change?* (New York, Harper & Row, 1978).

Harrison, W. C., *Dr. William Harvey and the Discovery of Circulation* (New York, MacMillan, 1967).

Harvey, W., *On the motion of the heart and blood in animals*, trans. Robert Willis, Great Minds – Science. Amherst, (New York, Prometheus Books, 1993).

Hay, S. I. (2001), 'The world of smoke, mirrors and climate change', *Trends Parasitol*, vol. 17, no. 10, p. 466.

Hay, S. I. *et al.* (2002), 'Climate change and the resurgence of malaria in the East African highlands', *Nature*, vol. 415, pp. 905–9.

Homer, *The Iliad*, trans. Robert Fagles (New York, Viking Penguin, 1990).

Houghton, J. T. *et al.* (eds), *The Science of Climate Change. Contribution of Working Group I to the Second Assessment of the Intergovernmental Panel on Climate Change (IPCC)*, (Cambridge, Cambridge University Press, 1996).

James, S. P. (1920), 'The disappearance of malaria from England', *Proc. R. Soc. Med.*, vol. 23, pp. 71–85.

Jetten, T. H. and Focks, D. A. (1997), 'Potential changes in the distribution of dengue transmission under climate warming', *American Journal of Tropical Medicine and Hygiene*, vol. 57, no. 3, pp. 285–97.

Jones, W. H. S., *Malaria and Greek History* (Manchester, Manchester University Press, 1909).

Jones, W. H. S., *Malaria: a neglected factor in the history of Greece and Rome* (Cambridge, Macmillan and Bowes, 1907).

Karl, T. R. *et al.* (1996), 'Indices of climate change in the United States', *Bulletin of the American Meteorological Society*, vol. 77, pp. 279–92.

Kerr, R. A. (1997), 'Greenhouse forecasting still cloudy', *Science*, vol. 276, pp. 1040–42.

Kovats, R. S. *et al.* (1999), 'Climate change and human health in Europe', *British Medical Journal*, vol. 318, pp. 1682–85.

Laigret, M. (1953), 'Remarques d'ordre général concernant la régression du paludisme dans nos pays', *Revue Path.* hum. comp. vol. 53, pp. 1308–12.

Lamb, H. H., *Climate, History and the Modern World* (London, Routledge, 1995).

Langholf, V., *Medical theories in Hippocrates: early texts and the 'Epidemics', Untersuchungen zur Antiken Literature und Geschichete, Band 34* (Berlin, Walter de Gruyter, 1990).

Levi-Strauss, C., *Tristes Tropiques* (Penguin, USA, 1992).

Lindsay, S. W. and Birley, M. H. (1996), 'Climate change and malaria transmission', *Ann. Trop. Med. Parasitol.*, vol. 90, no .6, pp. 573–88.

Lindzen, R. S. (1997), 'Can increasing carbon dioxide cause climate change?' *Proceedings of the National Academy of Sciences*, vol. 94, pp. 8335–42.

Lloyd, C. and Coulter, J. L. S., *Medicine and the Navy*, vol. 1. (Edinburgh, Livingstone, 1961).

Macdonald, G., *The Epidemiology and Control of Malaria* (Oxford, Oxford University Press, 1957).

Manning, M. and Nobre, C. (eds) (2001), *Technical Summary. Climate Change 2001: Impacts, Adaptation and Vulnerability. Contribution of Working Group II to the Third Assessment Report of the Intergovernmental Panel on Climate Change (IPCC)* (Cambridge, Cambridge University Press, 2001).

Martens, P., *Health and Climate change: Modelling the Impacts of Global Warming and Ozone Depletion* (London: Earthscan Publications, 1998).

McMichael, A. J., Patz, J. and Kovats, R. S., 'Impacts of Global Environmental Change on Future Health and Health Care in Tropical Countries', *British Medical Bulletin*, vol. 54, no. 2, pp. 475–88.

McMichael, A. J., *et al.*, *Climate Change and Human Health* (Geneva, World Health Organization, 1996).

Michaels, P. J. and Knappenberger, P. C. (1996), 'Human effect on global climate?' *Nature*, vol. 384, pp. 522–23.

Michaels, P. J. and Balling, R. C., *The Satanic Gases* (Washington, DC, Cato Institute, 2000).

Mouchet, J., *et al.* (1998), 'Evolution of Malaria in Africa for the Past 40 Years:

Impact of Climatic and Human Factors', *Journal of the American Mosquito Control Association*, vol. 14, no. 2, pp. 121–30.

Molineaux, L. and Gramiccia, G., *The Garki Project: Research on the Epidemiology and Control of Malaria in the Sudan Savanna of West Africa* (Geneva: World Health Organization, 1980).

Molineaux, L. (1985), 'The Pros and Cons of Modelling Malaria Transmission', *Trans R. Soc. Trop. Med. Hyg.*, vol. 79, no. 6, pp. 743–47.

Molineaux, L. and Dietz, K. (1999), 'Review of Intra-Host Models of Malaria', *Parassitologia*, vol. 41, nos. 1–3, pp. 221–31.

Patz, J. A. *et al.* (1996), 'Global Climate Change and Emerging Infectious diseases', *Journal of the American Medical Association*, vol. 275, no. 3, pp. 217–23.

Patz, J. A. *et al.* (1998), 'Dengue Fever Epidemic Potential as Projected by General Circulation Models of Global Climate Change', *Environmental Health Perspectives*, vol. 106, no. 3, pp. 147–53.

Ponte, L., *The Cooling* (Englewood Cliffs, Prentice-Hall, 1976).

Poser, C. M. and Bruyn, G. W., *An Illustrated History of Malaria* (New York, Parthenon, 1999).

Rampino, M., *Climate – History, Periodicity and Predictability* (Van Nostrand Reinhold, 1987).

Reiter, P. (2000), 'From Shakespeare to Defoe: Malaria in England in the Little Ice Age', *Emerging Infectious Diseases*, vol. 6, no. 1, pp. 1–11.

Reiter, P. (2001), 'Climate Change and Mosquito-Borne Disease', *Environmental Health Perspectives*, vol. 109, Suppl. 1, pp. 141–61.

Renkonen, K.O. (1944), 'Über das Vorkommen von Malaria in Finnland', *Acta Medica Scandinavica*, vol. 119, pp. 261–75.

Rogers, D. J. and Randolph, S. E. (2000), 'The global spread of malaria in a future, warmer world', *Science*, vol. 289, pp. 1763–6.

Ross, A. *Strange weather. Culture, Science, and Technology in the Age of Limits* (London, Verso, 1991).

Russell, P. F. (1956), 'World-wide malaria distribution, prevalence, and control', *American Journal of Tropical Medicine and Hygiene*, vol. 5, pp. 937–65.

Sabatinelli, G. (1998), 'Malaria Situation and Implementation of the Global Malaria Control Strategy in the WHO European region', World Health Organization Expert Committee on Malaria, 1998. MAL/EC20/98.9.

Shanks, G. D. *et al.* (2000), 'Changing Patterns of Clinical Malaria Since 1965 Among a Tea Estate Population Located in the Kenyan Highlands', *Trans. R. Soc. Trop. Med. Hyg.*, vol. 94, pp. 253–55.

Siegel, R. E. and Poynter, F. N. L. (1962), 'Robert Talbor, Charles II and Cinchona. A Contemporary Document', *Medical History*, vol. 6, pp. 82–85.

Simic, C. (1956), 'Le Paludisme en Yougoslavie', *Bulletin of the World Health Organization*, vol. 15, pp. 753–66.

Tett, S. F. B. *et al.* (1999), 'Causes of Twentieth-Century Temperature Change Near the Earth's Surface', *Nature*, vol. 399, pp. 569–72.

Watson, R. T., Zinyowera, M. C. and Moss, R. H. (eds), *Impacts, Adaptations and Mitigation of Climate Change: Scientific-Technical analyses. Contribution of Working Group II to the Second Assessment of the Intergovernmental Panel on Climate Change* (Cambridge, Cambridge University Press, 1996).

Wesenberg-Lund, C. (1921), 'Sur les Causes du Changement Intervenu dans le Mode de Nourriture de l'Anophèle Maculipennis', *C. r. Séanc. Soc. Biol.*, vol. 85, pp. 383–86.

World Health Organization (WHO) (1978), 'The Malaria Situation in 1976', *WHO Chronicles*, vol. 32, pp. 9–17.

Wigley, T. M. L., Ingram, M.J. and Farmer, G. (eds), *Climate and History* (Cambridge, Cambridge University Press, 1981).

Wigley, T. M. L. and Schimel, D. (eds) (2000), *The Carbon Cycle* (Cambridge, Cambridge University Press, 2000).

Zucker, J. R. (1996), 'Changing Patterns of Autochthonous Malaria Transmission in the United States: a Review of Recent Outbreaks', *Emerging Infectious Diseases*, vol. 2, no. 1, pp. 37–89.

2 Barriers to barriers: why environmental precaution has delayed mobile floodgates to protect Venice

Agenzia Giornalistica Italia, 'MOSE: Galan, there is hope for Venice', http://www.agi.it, 5 April 2003.

Ammerman, A. J., McClennen, C. E., De Min, M. and Housley, R. (1999), 'Sealevel change and the archaeology of early Venice', *Antiquity* 73, pp. 303–12.

Ammerman, A. J. and McClennen, C. E., 'Saving Venice', *Science*, vol. 289, issue 5483, p. 1301, 25 September 2000.

Anon, 'Italy Starts Work on Sea Gates Intended to Avert Venice Floods', *New York Times*, 15 May 2003.

Baccaro, A., 'Dighe mobili a Venezia. Via al piano dopo 37 anni', *Corriere della Sera*, 4 April 2003.

Berstein, P., *Against the Gods. The Remarkable Story of Risk* (New York, John Wiley, 1998), pp. 48–50.

Biorcio, R. (2002), 'Green Parties in National Governments: Italy', *Environmental Politics*, Spring, vol. 11, no. 1.

Brotto, M. T. and Gentilomo, M. (1998), 'The Venice Lagoon Project. The Barriers at the Lagoon Inlets for Controlling High Tides', in *Bulletin AIPCN-PIANC*, no. 98.

Cecconi, G., Canestrelli, P., Corte, C. and Di Donato, M. (1997a), 'Climate Record of Storm Surges in Venice', in *Impact of Climate Change on Flooding and Sustainable River Management*, RIBAMOD Workshop, Wallingford, 26–27 February 1998.

Cecconi, G. (1997b), 'Real time storm and watershed inflow forecasting in the Venice lagoon', in *Integrated System for Real Time Flood Forecasting and Warning*, RIBAMOD Workshop, Monselice, 25 September 1997.

Cecconi, G. (1997c), 'The Venice Lagoon Mobile Barriers. Sea Level Rise and Impact of Barrier Closures', in *Italian Days of Costal Engineering*, 16 May 1997.

Connor, S., 'Venice flood barriers scheme "will soon be obsolete" ', *Independent*, 13 May 2002.

Day, J. W., Are, D., Rismondo, A., Scarton, F. and Cecconi, G. (1996), 'Relative

Sea Level Rise and Venice Lagoon Wetlands', in *Active Protection and Water Flow Restoration of the Venice Lagoon*, Monographic Supplement to Quaderni Trimestrali.

Day, JR. J. W. *et al.* (1999), 'Soil Accretionary Dynamics, Sea Level Rise and the Survival of Wetlands in the Venice Lagoon: a field and modelling Approach', in *Estuarine, Coastal and Shelf Science*, 49.

Fay, S. and Knightley, P., *The Death of Venice* (New York, Praeger Publishers, 1976).

Fornasier, C. 'MOSE, via libera dopo 37 anni di odissea', *Corriere del Veneto*, 4 April 2003.

Gentilomo, M. and Cecconi, G. (1995), 'The Venice Barriers', in *Spatial Structures: Heritage, Present and Future*, International Symposium, Milan/Padua, vol. 1.

Gentilomo, M. and Warnock, J. (1997), *Sustainable Safeguards for the Protection of the Lagoon and the City of Venice*, Atti del IX World Water Congress, Montreal, Canada.

Gentilomo, M. (1997), *The Venice Lagoon Project. Mobile barriers at the lagoon inlets for controlling high tides*, in Atti del Symposium del 24 maggio 1997, Florence International Conference on Large Dams (ICOLD).

Gentilomo, M. *et al.* (1999), *Environmental Management Framework for Ports and Related Industries*, International Navigation Association, Report of Working Group 4.

Ghetti, A. (1988), 'Subsidence and Sea-Level Fluctuations in the Territory of Venice', in *Landscape and Urban Planning*, no. 16.

Goklany, I., *The Precautionary Principle. A Critical Appraisal of Environmental Risk Assessment* (Washington DC, Cato Institute, 2001).

Guthrie, R. (2001), 'History of Venice on the Water', Save Venice Inc., http://www.savevenice.org/, 22 January 2001.

International College of Experts (1998), 'Report on the mobile gates project for the tidal flow regulation at the Venice lagoon inlets', Venice, in *Numero speciale dei Quaderni Trimestrali*.

The Italian Ministry of Public Works, The Venice Water Authority and the New Venice Consortium (1997), *Environmental Impact Study (EIS) of the Preliminary Plan for Mobile Barriers at Lagoon Inlets for the Defence Against High Waters* (Venice).

Italia Nostra, Venice Chapter, 'A Commitment for Action', http://www.provincia.venezia.it/italianostra/1quale/1which.htm

James, B., 'A global threat laps at the gates of Venice', the *International Herald Tribune*, 22 March 2002.

Keahey, J., *Venice Against the Sea* (New York, St Martins Press; 2002).

Lomborg, B., *The Skeptical Environmentalist. Measuring the Real State of the World* (Cambridge, UK, Cambridge University Press, 2001).

Mazzacurati, G. (1996), 'Water Flow Restoration and the Active Protection of the Venice Lagoon', in *Active Protection and Water Flow Restoration of the Venice Lagoon*, Monographic Supplement to Quaderni Trimestrali.

Morris, J., *Venice,* (London, Faber and Faber, 1993).

Norwich, J. J., *A History of Venice* (London, Penguin, 1982).

Podger, C., 'Floodgates "won't save Venice" ', BBC News Europe Online, 13 May 2002.

Pirazzoli, P. (1992), 'Did the Italian Government Approve an Obsolete Project to Save Venice?' *EOS*, Transactions, American Geophysical Union, vol. 83, no. 20, 14. 5. 2002, pp. 217–23.

Runca, E., Bernstein, A., Postma, L. and Di Silvio, G. (1996), 'The framework of analysis to evaluate environmental measures in the Venice lagoon', in *Active Protection and Water Flow Restoration of the Venice Lagoon*, Monographic Supplement to Quaderni Trimestrali.

Scarton, F., Perco, F. and Borella, S. (1996), 'The importance of protecting lagoon habitats for bird life', in *Active Protection and Water Flow Restoration of the Venice Lagoon*, Monographic Supplement to Quaderni Trimestrali.

Standish, D. (2001a), 'Beware of the Precautionary Principle', Italy Daily section of the *International Herald Tribune*, 24 April 2001.

Standish, D. (2001b), 'Flap Project Will Save Venice', Italy Daily section of the *International Herald Tribune*, 10 December 2001.

Standish, D. (2002a), 'Big projects, small minds', *Spiked-online*, http://www.spiked-online.com, 17 July 2002.

Standish, D. (2002b), 'Will sinking Venice raise our ambitions?' Italy Daily section of the *International Herald Tribune*, 3 December 2002.

Standish, D. (2003a), 'A view from the bridge', in *The Architects' Journal*, 3 April 2003, pp. 38–39.

Standish, D. (2003b), 'Barriers to Barriers', Tech Central Station, http://www.techcentralstation.be/2051/, 29 April 2003.

Vitucci, A., 'Berlusconi arriverà il 29 aprile', *La Tribuna di Treviso*, 4 April 2003.

World Wildlife Fund (WWF Italy), 'MOSE, WWF e Italia Nostra ricorrono al TAR', http://www.wwf.it/news/322003_1634.asp, 3 February 2003.

Woodard, C., 'The Sinking City', *E/The Environment Magazine*, vol. XIV, no. 2, March–April 2003, http://www. emagazine.com/march-april_2003/0303curr_venice_intro.html

Zamparutti, T., 'Venice in Peril', *The Ecologist*, 22 March 2003.

Zitelli, A. and Rossetto, P. (1996), 'Techniques in the environmental restoration programme of the Venice lagoon: The 'Palude della Rosa' pilot project', in *Active Protection and Water Flow Restoration of the Venice Lagoon*, Monographic Supplement to Quaderni Trimestrali.

Zuccchetta, G. (2000), *Storia dell'acqua alta a Venezia*, Marsilio, Venice.

3 Climate change: the 21st century's most urgent environmental problem or proverbial last straw?

Arnell, N. W. (1999), 'Climate Change and Water Resources,' *Global Environmental Change* 9: S31–49.

Arnell N.W. *et al.* (2002). 'The Consequences of CO^2 Stabilization for the Impacts of Climate Change,' *Climatic Change* 53, pp. 413–46.

Ausubel, J. H. (1991), 'Does Climate Still Matter?', *Nature* 350, pp. 649–52.

Byerlee, D. and Echeverria, R. G. (eds) (2002). *Agricultural Research Policy in an*

Era of Privatization (CABI International). Chapters 1&2 online at http://
www.cabi-publishing.org/Bookshop/ReadingRoom/0851996000.asp, visited 9
July 2003.

Clinton, W. J., *State of the Union Address* (1998), online at
http://www.usemb.ee/union98.php3, visited 7 July 2003.

Energy Information Administration (EIA) (1998), 'What Does the Kyoto Protocol
Mean to U.S. Energy Markets and the U.S. Economy?' SR/OIAF/98-03(S),
Washington: EIA.

Food and Agricultural Organization (FAO) (2001), *Global Forest Resources
Assessment 2000* (Rome: FAO).

FAO (1996), *The State of Agriculture* (Rome, Italy: FAO).

FAO (2002), *State of Food Insecurity in the World 2002* (Rome: FAO).

FAOSTAT Database, online at http://apps.fao.org, visited 30 June 2003.

Fitter, A. H. and Fitter, R. S. R. (2002), 'Rapid Changes in Flowering Time in
British Plants,' *Science* 296, pp. 1689–91.

Fox, R. (2001a), *Millennium Atlas of Butterflies in Britain and Ireland*, Butterfly
Conservation press release, 2 March, online at www.butterfly-
conservation.org/ne/news/bnm/english.html, visited 8 December 2002.

Fox, R. (2001b), *A Butterfly's Map of Climate Change*, 9 October 2001, online at
www.changingclimate.org/content/articles/article/data/section_3/article_7/
part_18/, visited 8 December 2002.

Intergovernmental Panel on Climate Change (IPCC) (2001a), *Climate Change
2001: The Scientific Basis* (New York, Cambridge University Press, 2001).

IPCC (2001b), *Climate Change 2001: Mitigation* (Cambridge [UK], Cambridge
University Press, 2001).

Gitay, H. *et al.* (2001), 'Ecosystems and their Goods and Services,' in IPCC,
Climate Change 2001: Impacts, Adaptation, and Vulnerability (New York,
Cambridge University Press, 2001).

Goklany, I. M. (1992), 'Adaptation and Climate Change', prepared for 1992
Annual Meeting of the American Association for the Advancement of Science
(Chicago IL, 6–11 February 1992).

Goklany, I. M. (1995), 'Strategies to Enhance Adaptability: Technological
Change, Sustainable Growth and Free Trade', *Climatic Change* 30,
pp. 427–49.

Goklany, I. M. (1998), 'Saving Habitat and Conserving Biodiversity on a
Crowded Planet', *BioScience* 48, pp. 941–53.

Goklany, I. M. (1999a), 'Meeting Global Food Needs: the Environmental Trade-
offs Between Increasing Land Conversion and Land Productivity', *Technology*
6, pp. 107–30.

Goklany I. M. (1999b), 'Richer Is More Resilient: Dealing with Climate Change
and More Urgent Environmental Problems', in R. Bailey (ed.), *Earth Report
2000: The True State of the Planet Revisited* (New York, McGraw-Hill, 1999).

Goklany, I. M. (2000), 'Potential Consequences of Increasing Atmospheric CO_2
Concentration Compared to Other Environmental Problems', *Technology* 7S,
pp. 189–213.

Goklany, I. M. (2001a), *Economic Growth and the State of Humanity* (Bozeman,
MT: Political Economy Research Center).

Goklany, I. M. (2001b), 'The Problem of the Last Straw: The Case of Global
   Warming', in Dorf, R. (ed.), *Technology, Humans, and Society: Toward a
   Sustainable World* (San Diego, Academic Press, 2001).
Goklany, I. M. (2002a), 'Affluence, Technology, and Well-Being', *Case Western
   Reserve Law Review* 53, pp. 369–90.
Goklany, I. M. (2002b), 'Comparing 20th Century Trends in U.S. and Global
   Agricultural Land and Water Use', *Water International* 27, pp. 321–29.
Goklany, I. M. (2002c), 'The Globalization of Human Well-being', *Policy
   Analysis*, no. 447, Cato Institute.
Goklany, I. M. (2003), 'Relative Contributions of Global Warming to Various
   Climate Sensitive Risks, and Their Implications,' (in review; paper submitted
   to journal for review).
Greenwire (1998), *Worldview – Climate Change II: Scientists Fear Warming*
   (cited 5 February 1998), available through search engine at
   www.nationaljournal.com/pubs/greenwire/extra/search.htm.
Ha-Duong, M., Grubb, M. J. and Hourcade, J. C. (1997), 'Influence of
   Socioeconomic Inertia and Uncertainty on Optimal Co^2-Emission Abatement',
   Nature 390, pp. 270–73.
Henderson-Sellers, A. *et al.* (1998a), 'Tropical Cyclones and Global Climate
   Change: a Post-IPCC Assessment', *Bulletin of the American Meteorological
   Society* 79, pp. 19–38.
Henderson-Sellers, A. (1998b), 'Climate Whispers: Media Communication about
   Climate Change', *Climatic Change* 40, pp. 421–56.
Hulme, M., Mitchell, J., Ingram, W., Lowe, J., Johns, T., New, M. and Viner, D.
   (1999), 'Climate Change Scenarios for Global Impact Studies', *Global
   Environmental Change* 9, pp. S3–S19.
Knowlton, Nancy (2000), 'The Future of Coral Reefs', in *The Future of
   Evolution*, National Academy of Sciences Colloquium.
McMichael A. *et al.* (2001), 'Human Health', in IPCC (2001b).
McNeely, J. A. *et al.* (1995), 'Human influences in biodiversity', in V. H.
   Heywood, *et al.* (eds), *Global Biodiversity Assessment* (Cambridge [UK],
   Cambridge University Press, 1995), pp. 755–57.
Malakoff, D. (1997), 'Thirty Kyotos Needed to Control Warming', *Science* 278,
   p. 2048.
Myneni, R. B. *et al.* (1997), 'Increased Plant Growth in the Northern High
   Latitudes', *Nature* 386, pp. 698–702.
Nicholls, N. (1997), 'Increased Australian Wheat Yield Due to Recent Climate
   Trends', *Nature* 387, pp. 484–85.
Parmesan C. and Yohe, G. (2003), 'A Globally Coherent Fingerprint of Climate
   Change Impacts Across Natural Systems', *Nature* 421, pp. 37–42.
Parry, M. L., Rosenzweig, C., Iglesias, A. Fischer, G. and Livermore, M. (1999),
   'Climate Change and World Food Security', *Global Environmental Change* 9,
   pp. S51–S67.
Pearce, D. W., *et al.* (1996), 'The Social Costs of Climate Change: Greenhouse
   Damage and the Benefits of Control', in IPCC, *Climate Change 1995:
   Economic and Social Dimensions of Climate Change* (New York, Cambridge
   University Press, 1995).

Root, T. L., Price, J. T., Hall, K. R., Schneider, S. H., Rosenzweig, C. and Pounds, J. A. (2003), 'Fingerprints of Global Warming on Wild Animals and Plants', *Nature* 421, pp. 57–60.

Royal Society for the Protection of Birds (RSPB) (2000), *The State of the UK's Birds 2000*.

Royal Society for the Protection of Birds (RSPB) (2001), *The State of the UK's Birds 2001*.

Wildlife News (2002), 'Climate change turns up the heat on UK's threatened birds,' 10 August 2002, visited 8 December 2002.

Solomon, A. F. *et al.* (1996), 'Wood Production under Changing Climate and Land Use', in IPCC, *Climate Change 1995: Impacts, Adaptations and Mitigation* (Cambridge, Cambridge University Press, 1996).

Trenberth, K. (undated), 'Does Climate Change Cause More Extreme Weather? Yes', in *Washington Post*, Issues from Global Climate Change, undated, online at www.washingtonpost.com/wp-adv/specialsales/nei/global/article9.htm, visited 4 December 2002 (no longer available).

Vaughan, D. G., Marshall, G. J., Connolley, W. M, King, J. C. and Mulvaney, R. (2001), 'Climate Change: Devil in the Detail', *Science* 293, pp. 1777–79.

Walther, G. R. *et al.* (2002), 'Ecological Responses to Recent Climate Change', *Nature* 416. pp. 389–95.

Wigley, T. M. L. *et al.* (1996), 'Economic and Environmental Choices in the Stabilization of Atmospheric CO^2 Concentrations', *Nature* 379, pp. 240–43.

Wigley, T. M. L. (1997), 'Implications of Recent CO_2 Emission-Limitation Proposals for Stabilization of Atmospheric Concentrations', *Nature* 390, pp. 267–70.

Wigley, T. M. L. (1998), 'The Kyoto Protocol: CO_2, CH_4 and Climate Implications', *Geophysical Research Letters* 25, pp. 2285–88.

Woodwell, G. (1997), 'Exaggeration or underestimate', *Nature* 390, p. 547.

World Bank, *World Development Indicators 2002*.

World Bank, *World Development Indicators*, CD-ROM (Washington DC, World Bank, 2002).

World Health Organization (1999), *World Health Report 1999* (Geneva: WHO).

WWF Finland (2002), *Climatic Change Has Altered Finnish Flora and Fauna*, press release, 17 May 2002, on file with author.

4 Is Kyoto a good idea?

Ågerup, M. (1998), *Dommedag er aflyst*, Gyldendal.

Balling, R. C., *The Heated Debate – Greenhouse Predictions Versus Climate Reality*, (Pacific Research Institute for Public Policy, San Francisco, 1992).

Bove, M. C. *et. al.* (1998), 'Effect of El Nino on US landfalling hurricanes, revisited', Bulletin of the *American Meteorological Society* 79 (11):2, 477–82.

Castles, I. and Henderson, D. (2003), 'Correspondence with the SRES', available at http://www.economist.com/displaygeneric.cfm?pageheadgif=FinanceandEconomics&key=efhp1 (last visited 12 July 2003).

Castles, I., and Henderson, D. (2003), 'The IPCC Emission Scenarios: An economic-statistical critique', *Energy and the Environment*, vol. 14, nos. 2&3, pp. 159–85.

Chagnon, S. A. *et al.* (2000), 'Human factors explain the increased losses from weather and climate extremes', *Bulletin of the American Meteorological Society* 81(3), pp. 437–42.

Chakravorty, U. and Tse, K.-P. (no date), 'Transition from Fossil Fuels to Renewable Energy', http://www. earthinstitute.columbia.edu/events/econSeminar/Chakravorty.pdf

Commission of the European Communities (2002), COM 702. Available at: http://europa.eu.int/eur-lex/en/com/rpt/2002/com2002_0702en01.pdf (last visited 13 July 2003).

Corcoran, T. (2002), 'An "insult to science" ', *National Post*, 14 December 2002. Available at: http://www.nationalpost.com/financialpost/story.html?id= %7B54B082F6-100C-4ADA-9012-EDF7EC30DD03%7D

Den Elzen, M. G. J., and Both, S. (2002), 'Modelling emissions trading and abatement costs in FAIR 1. 1 – Case study: The Kyoto Protocol under the Bonn-Marrakesh Agreement', RIVM (National Institute of Public Health and the Environment, Denmark) report 728001021, available at http://www.rivm.nl/bibliotheek/rapporten/728001021.html

The Economist, 'Hot Potato', 13 February 2003.

Eurostat (2002), Yearbook, Brussels, European Commission.

Fernandez, M. D., Pieters, A., Donoso, C., Tezara, W., Azuke, M., Herrera, C., Rengifo, E. and Herrera, A. (1998), 'Effects of a natural source of very high CO_2 concentration on the leaf gas exchange, xylem water potential and stomatal characteristics of plants of *Spatiphylum cannifolium* and *Bauhinia multinervia*', *New Phytologist* 138, pp. 689–97.

Friis-Christensen, E. and Lassen, K. (1991), 'Length of the Solar Cycle: An Indicator of Solar Activity closely Associated with Climate', *Science* 254, pp. 698–700 (1 November 1991).

Goklany, I. M. (2000), 'Potential Consequences of Increasing Atmospheric CO_2 Concentration Compared to other Environmental Problems', *Technology* vol. 7S, pp. 189–213.

Gouk, S. S., He, J. and Hew, C. S. (1999), 'Changes in photosynthetic capability and carbohydrate production in an epiphytic CAM orchid plantlet exposed to super-elevated CO_2', *Environmental and Experimental Botany* 41, pp. 219–30.

Gray, V., *The Greenhouse Delusion* (Brentwood, Essex, UK, Multi-Science Publishing Co. Ltd., 2002).

Grubb, M., 'The Economics of the Kyoto Protocol' in Owen, A. D. (ed.) *The Economics of the Kyoto Protocol* (forthcoming, details on file with author).

Henderson, D. (2003), 'SRES and IPCC: Further Concerns', available at http://www.sdnetwork.net/pdfs/ipcc_document3.pdf, pp. 7–10.

International Energy Agency (IEA) (2002), 'Beyond Kyoto: Energy Dynamics and Climate Stabilisation', Paris, OECD.

IEA (2002), *Key World Energy Statistics*, http://www.iea.org/statist/ keyworld2002/key2002/keystats.htm

Intergovernmental Panel on Climate Change (IPCC) (1996), *Climate Change*

1995 – Scientific-Technical Analyses of Impacts, Adaptations and Mitigations of Climate Change. Report of IPCC Working Group II (Cambridge, UK, Cambridge University Press, 1996).

IPCC (1998), *The Regional Impacts of Climate Change: An Assessment of Vulnerability*.

IPCC (2000), 'Special Report on Emissions Scenarios, Summary for Policymakers', http://www.ipcc.ch/pub/sres-e.pdf

IPCC (2001a), *Climate Change 2001: Synthesis Report – Summary for Policymakers*, available at http://www.grida.no/climate/ipcc_tar/vol4/

IPCC (2001b), *The Scientific Basis: Summary for Policymakers*, Working Group I, available at http://www.grida.no/climate/ipcc_tar/wg1/005.htm

IPCC (2001c), *The Scientific Basis: Technical Summary*, Climate Change 2001, Working Group I, available at http://www.grida.no/climate/ipcc_tar/wg1/010.htm

IPCC (2001d), *The Scientific Basis*, Climate Change 2001, Working Group I, available at http://www.grida.no/climate/ipcc_tar/wg1/index.htm

IPCC (2001e), *The Scientific basis: Executive Summary*, Climate Change 2001, Chapter 14, available at http://www.grida.no/climate/ipcc_tar/wg1/501.htm

IPCC (2001f), *Impacts, Adaptation and Vulnerability*, Climate Change 2001, Working Group II, available at http://www.grida.no/climate/ipcc_tar/wg2/index.htm

IPCC (2001g), *Mitigation*, Climate Change 2001, Working Group III, available at http://www.grida.no/climate/ipcc_tar/wg3/index.htm

Keatinge, W. R. *et al.* (2000), 'Heat related mortality in warm and cold regions of Europe: observational study', *British Medical Journal* 321 (7262), pp. 670–3. Available at: http://bmj.com/cgi/reprint/321/7262/670.pdf (last visited 13 July 2003).

Kristoffersen, A. (2002), *Danmarks omkostninger ved reduction af CO₂*, Institute for Environmental Assessment.

Kunkel *et al.* (1999), 'Temporal fluctuations in weather and climate extremes that cause economic and human health impacts: a review', *Bulletin of the American Meteorological Society* 80(6), pp. 1077–98.

Landes, D., *The Wealth and Poverty of Nations* (London, Abacus, 1998).

Laut, P. and Gundermann, J. (1998), 'Solar cycle length hypothesis appears to support the IPCC on global warming', *Journal of Atmospheric And Solar-Terrestrial Physics* 60, pp. 1719–28.

Lindzen, R. S., Chou, M-D., Hou, A. Y. (2001) 'Does the Earth have an adaptive infrared iris?' *Bulletin of the American Meteorological Society* 82, pp. 417–32.

Lomborg, B., *The Skeptical Environmentalist. Measuring the Real State of the World* (Cambridge, Cambridge University Press, 2001).

McKibbin, W. J. and Wilcoxen, P. J. (2003), 'Estimates of the Costs of Kyoto–Marrakesh Versus The McKibbin–Wilcoxen Blueprint', Brookings Institute (revised 24 February).

McKibbin, W. J. and Wilcoxen, P. J. (1999), 'Designing a Realistic Climate Change Policy that includes Developing Countries', 20 October. Paper prepared for the UN University Symposium on 'Global Environment and

Economic Theory', Tokyo, Japan, 24–25 October 1999. Available at: http://www.msgpl.com.au/msgpl/download/developing.pdf

McKitrick, R. (2003), 'Emission Scenarios & Recent Global Warming Projections', Fraser Forum, January.

Magné, B. and Moreaux, M. (2002), 'Long Run Energy Trajectories: Assessing the Nuclear Option In Response To Global Warming' (preliminary draft), February.

Manne, A. S. and Richels, R. G. (2001), 'US Rejection of the Kyoto Protocol: the impact on compliance costs and CO_2 emissions', paper presented to the Stanford University Energy Modeling Forum, 6 August 2001. Available at: http://www.stanford.edu/group/MERGE/kyoto.pdf

Marsh, N. and Svensmark, H. (2000). 'Cosmic Rays, Clouds and Climate', *Space Science Review* 94, pp. 215–30.

Michaels, P. J., 'The Greenhouse Effect and Global Change', in Julian Simon (ed.), *The State of Humanity* (Oxford, Blackwell Publishers, 1995).

Nakicenovic, N. and Riahi, K. (2002), 'An Assessment of Technology Change', International Institute for Applied Systems Analysis. Available at: http://www.iiasa.ac.at/Publications/Documents/RR-02-005.pdf.

National Academy of Sciences (NAS) (2001), *Climate Change Science: An analysis of some key questions*, (Washington, DC, National Academy Press, 2001). Available at: http://www.nap.edu/html/climatechange/

Nordhaus, W. D. (2001), 'Global Warming Economics', *Science*. 294, pp. 1283–4. (9 November).

Nordhaus, W. D. and Boyer, J. G. (1999), 'Requiem for Kyoto: An Economic Analysis of the Kyoto Protocol', (8 February). Available at: http://www.econ.yale.edu/~nordhaus/homepage/Kyoto.pdf

Nordhaus, W. D. and Boyer, J. G. (1999), 'Roll the Dice Again: Economic Models of Global Warming', MIT Press (25 October), internet version: http://www.econ.yale.edu/%7Enordhaus/homepage/dice_section_I.html

OECD (2003), *Main Economic Indicators*

Pielke, R. A. Jnr and Landsea, C. W. (1998), 'Normalized hurricane damages in the United States: 1925–1995', *Weather and Forecasting* 13(3), pp. 621–31.

Rosenzweig, C. and Parry, M. L. (1994), 'Potential impact of climate change on world food supply', *Nature* 367, pp. 133–38.

Smith, F. L., 'Conclusion: The Role of Opportunity Costs in the Global Warming Debate', in Adler, J. H., *The Costs of Kyoto* (Washington, DC, Competitive Enterprise Institute, 1997).

Titus, J. G. *et al.* (1991), Greenhouse Effect and Sea Level Rise: The Cost of Holding Back the Sea', *Coastal Management*, vol. 19, pp. 171–204. Available at: http://yosemite.epa.gov/oar/globalwarming.nsf/content/ResourceCenterPublicationsSLRCost_of_Holding.html (last visited 13 July 2003).

UNDP (2002), *Human Development Report* (Oxford, Oxford University Press, 2002). Available at: http://hdr.undp.org/reports/global/2002/en/

UNPP (2002), *World Population Prospects: The 2002 Revision*. Available at http://esa.un.org/unpp/

Webster, *et al.* (2001), 'Uncertainty in Emissions Projections for Climate Models',

MIT Joint Program on the Science and Policy of Global Change, Report No. 79 (August). Available at: http://www.mit.edu/afs/athena.mit.edu/ org/g/globalchange/www/MITJPSPGC_Rpt79.pdf (last visited 13 July 2003).

Weyant, J. P. and. Hill, J. N. (1999), 'Introduction and overview', in *The Costs of the Kyoto Protocol: A Multi-Model Evaluation*, special issue of *The Energy Journal*. See http://www.iaee.org/en/publications/kyoto.aspx

5 Sustainable energy for the poor

Aukland, L., Moura Costa, P., Bass, S., Huq, S., Landell-Mills, N., Tipper, R. and Carr, R. (2002). 'Laying the Foundations for Clean Development: Preparing the Land Use Sector. A quick guide to the Clean Development Mechanism', (London, IIED, 2002).

Bruce, N., Perez-Padilla, R. and Albalak, R. (2002), 'The health effects of indoor air pollution exposure in developing countries', Geneva, World Health Organization. Available at: http://www.who.int/peh/air/Indoor/ oeho205ari.htm

Department for International Development (DFID) (2002), 'Energy for the Poor', London, DFID. Available at: http://www.livelihoods.org/info/docs/ WSSD_Energy.pdf

Energy Information Agency (EIA) (2001), 'India: Environmental Issues', US Department of Energy (June). Available at: http://www.eia.doe.gov/emeu/cabs/ indiaenv.html

Goldemberg, J. and Johansson, T. B. (1995), 'Energy as an Instrument for Socio-Economic Development', Energy and Atmosphere Programme, United Nations Development Programme (New York, UNDP). Available at: http://www.undp.org/seed/energy/policy/overview.htm

Guru, Sutanu (2002). 'Renewable Energy Sources in India: Is it Viable?', Liberty Institute Working Paper (New Delhi, Liberty Institute). Available at: http://www.libertyindia.org/pdfs/renewable_energy_guru_october2002.pdf

International Energy Agency (IEA), *World Energy Outlook* (Paris, IEA, 2002).

Kyoto Protocol to the United Nations Framework Convention on Climate Change, Article 12, Paragraph 2. Available at: http://unfccc.int/resource/docs/ convkp/kpeng.pdf

Narain, Sunita (2003), 'Killing the high end killer'; *Down to Earth*, vol. 12, no. 4, 15 July 2003, p. 5.

Ravindranath, N. H. and Hall, D. O., *Biomass, Energy and Environment: A Developing Country Perspective from India* (Oxford, Oxford University Press, 1999).

Ravindranath, N. H., *Renewable Energy and Environment: A Policy Analysis for India* (New Delhi, Tata McGraw-Hill, 2000).

Reddy, B. S., 'Energy Efficiency and Environmental Implications in India's Household Sector', Indira Gandhi Institute of Development Research. Available at: http://www.rite.or.jp/GHGT6/pdf/H3-1.pdf

Smith, K. R. (2000), 'National burden of disease in India from indoor air pollution', *Proceedings of the National Academy of Sciences*, 21 November.

vol. 97, no. 24, pp. 13286–13293. Available at: http://www.who.int/
environmental_information/Disburden/wsh00-7/Methodan6-5.htm
Smith, K. R. and Mehta, S. (2000), 'Estimating the global burden of disease from
indoor air pollution', World Health Organization. Available at: www.who.int/
environmental_information/Disburden/wsh00-7/Methodan6-5.htm
Sobhani, L. and Retallack, S., 'Fuelling Climate Change', in Mander, J. and
Goldsmith, E. (eds) *The Case Against the Global Economy* (London,
Earthscan, 2001).
Subudh, R. N., *Energy Options for 21st Century* (New Delhi, Ashish Publishing
House, 1993).
Sukla, P. R. (1997), 'Biomass Energy in India: Policies and Prospects in Biomass
Energy'. Key Issues and Priority Needs, Conference Proceeding, Paris, France,
3–5 February, International Energy Agency.
Tata Energy Research Institute (2002), 'Integrating New and Sustainable
Technologies for Eliminating Poverty (INSTEP): Opportunities and Challenges
from Globalization'. Available at: http://www.terina.org/prog/11012002.htm
Toman, M. and Cazorla, M. (1998), 'About the Clean Development Mechanism:
A Primer', Resources for the Future, Washington, DC. Available at:
www.weathervane.rff.org/features/feature048.html
United Nations Development Programme (UNDP) (2002). 'Clean Energy for
Development and Economic Growth: Biomass and other Renewable Energy
Options to Meet Energy and Development Needs in Poor Nations', Policy
Discussion Paper for the Environmentally Sustainable Development Group
(ESDG) of the UNDP. Available at: http://www.undp.org/seed/eap/html/
publications/2002/2002b.htm
World Health Organization (WHO) (2000), 'Addressing the links between indoor
air pollution, household energy, and human health', Geneva, World Health
Organisation. Available at: http://www.who.int/mediacentre/events/
HSD_Plaq_10.pdf.
WHO (no date). 'An Anthology on Women, Health and Environment: Domestic
Fuel Shortage and Indoor Air Pollution'. Available at: http://www.who.int/
environmental_information/Women/womfuel.htm
World Bank, *World Development Indicators* (Washington, DC, World Bank, 2002).

6 Energy for the poor? The clean development mechanism

Calder, Nigel (1975), 'In the Grip of a New Ice Age', *International Wildlife*,
journal of the National Wildlife Federation, July.
Energy Research Institute (ERI) (2002), 'The Clean Development Mechanism',
University of Cape Town, South Africa. Available at: http://www.eri.uct.ac.za/
eri%20publications/CDM%20-%20A%20Guide%20for%
20Potential%20Participants%20in%20SA.pdf
Hirschberg, S., Spiekerman, G., and Dones, R. (1998), 'Severe Accidents in the
Energy Sector', PSI Report, nr. 98–16, Villigen-PSI.
Keigwin, L. D. (1996), 'The Little Ice Age and Medieval Warm Period in the
Sargasso Sea', *Science* 274, pp. 1504–8.

Lemonick, M. D., 'The Ice Age Cometh?' *Time*, 31 January 1994. Available at: www.time.com/time/archive/preview/from_redirect/0,10987,1101940131-163735,00.html

Lomborg, B., *The Skeptical Environmentalist. Measuring the Real State of the World* (Cambridge, UK, Cambridge University Press, 2001).

Nelson, R. H. (2003), 'Environmental Colonialism: Saving Africa from the Africans', *Independent Review*, vol. VIII, no. 1, Summer 2003, pp. 65–86. Available at: www.independent.org/tii/media/pdf/tir81nelson.pdf

Paraffin Safety Association of South Africa (2001), Annual Report.

Simon, J. L., *The Ultimate Resource* (Princeton, NJ, Princeton University Press, 1981).

United Nations Population Division (UNPD), *World Population Prospects* (New York, United Nations, 1998).

van de Vate, Joop F (1997), 'Comparison of energy sources in terms of their full energy chain emission factors of greenhouse gases', *Energy Policy*, vol. 25, no. 1. Elsevier, pp. 1–6.

7 Warming aid, chilling trade?

Appleton, A. E. (1999), 'Environmental Labelling Schemes: WTO Law and Developing Country Implications', in Sampson and Chambers, *Trade, Environment and the Millennium* (Tokyo, United Nations University Press, 1999), pp. 195–221.

Barrett, S. (1994), 'Self-Enforcing International Environmental Agreements', *Oxford Economic Papers*, vol. 46, pp. 878-94.

ECO (1997), 'OUTREACH 1997', vol. 1, no. 23, 21 April. Available at: http://habitat.igc.org/csd-97/0r-9723.html (accessed 13 July 2003).

Goldberg, D. (1994), 'GATT Tuna-Dolphin II: Environmental Protection Continues To Clash With Free Trade', *Center for International Environmental Law*, June, no. 2.

Hogue, Arthur R. (1966 [1985]). *Origins of the Common Law* (Indianapolis, Liberty).

Morris, J., *The Political Economy of Land Degradation* (London, Institute of Economic Affairs, 1995).

Morris, J., 'International Environmental Agreements: Developing Another Path', in *The Greening of U. S. Foreign Policy*, Anderson, T. L. and Miller, H. (eds). (Stanford, California: Hoover Institution Press, 2000), pp. 267–301.

Pearce, D. (1994), 'The Greening of the GATT: Some Economic Considerations,' in Cameron, Demaret and Geradin, *Trade and the Environment – The Search for Balance* (London, Cameron May Ltd, 1994). pp. 20–38.

't Sas-Rolfes, Michael, *Rhinos: Conservation, Economics and Trade-Offs* (London, Institute of Economic Affairs, 1995).

United States Council for International Business (USCIB) (2002), 'US Faces Trade Sanctions Over Kyoto Plan', 11 December. available at: http://www.uscib.org/%5Cindex.asp?documentID=2496 (accessed 17 July 2003).

United States – Measures Affecting Alcoholic and Malt Beverages (1992) Report of the Panel, GATT Document DS23/R, adopted 19 June.

United States – Import Prohibition of Certain Shrimp Products (1998), Report of the Appellate Body, WTO AB 1998-4.

Wildavsky, A., 'Foreword' to Balling, R. C., *The Heated Debate: Greenhouse Predictions versus Climate Reality* (San Francisco, Pacific Research Institute, 1992).

World Health Organization (WHO), *World Health Report* (Geneva, World Health Organization, 1995). For an overview, see: http://www.who.int/whr/2002/overview/en/print.html

8 How Europe's risk regulations affect business

Business in the Environment. *Investing in the Future — Survey of City attitudes to environmental and social issues in the UK* (London, Business in the Community, 2001).

Desrochers, Pierre, 'Does it pay to be green? Some historical perspectives', in Morris, Julian (ed.) *Sustainable Development: Promoting Progress or Perpetuating Poverty?* (London, Profile Books, 2002). Available at: http://www.sdnetwork.net/pdfs/pierre_desrochers_chapter3.pdf

DRI–WEFA (2002) *Kyoto Protocol and Beyond: The High Economic Cost to the United Kingdom.*

European Commission (1995) *An energy policy for the European Union*, COM (95)682).

European Commission (1997) *Energy for the Future: Renewable Sources of Energy*, COM (97)599. Available at http://europa.eu.int/comm/energy/library/599fi_en.pdf

European Commission (2003), 'Commissioner Wallström calls for more stringent measures and policies to cut EU greenhouse gas emissions', press release, 6 May. Available at: http://europa.eu.int/rapid/start/cgi/guesten.ksh?p_action.gettxt=gt&doc=IP/03/632|0|RAPID&lg=EN

European Environment Agency (EEA) (2002). "Greenhouse gas emission trends and predictions in Europe', *Environmental Issue Report* no. 33. Available at: http://reports.eea.eu.int/report_2002_1205_091750/index_html

EEA (2003), 'Annual European Greenhouse Gas Inventory 1990–2001 and Inventory report 2003', Technical Report no 95. Available at: http://reports.eea.eu.int/technical_report_2003_95/index_html

Friends of the Earth International (FOEI) (2001), "Towards binding corporate accountability", 5 October. Available at: http://www.foei.org/publications/corporates/accountability.html

International Council for Capital Formation (ICCF) (2002), "Kyoto Protocol and Beyond: Impacts on European Countries", October 2002. Available at: http://www.iccfglobal.org/PDFs/White%20Paper%20REV%2010-02.pdf

Lomborg, B. (2001) *The Skeptical Environmentalist. Measuring the Real State of the World* (Cambridge, UK, Cambridge University Press, 2001).

Macalister, T. (2003), "High price tag is put on green energy', *Guardian* (UK), 5 June. Available at: http://www.guardian.co.uk/business/story/0,3604,970497,00.html

Nordhaus, W. and Boyer, J. (1999), 'Requiem for Kyoto: An economic analysis of the Kyoto Protocol', *The Energy Journal: Kyoto* special issue, pp.93–130. Available at http://www.econ.yale.edu/~nordhaus/homepage/Kyoto.pdf

Parry, M., Rosenzweig, C., Iglesias, A., Fischer, G. and Livermore, M. (1998), 'Buenos Aires and Kyoto targets do little to reduce climate change impacts', *Global Environmental Change* 8, vol. 4, pp.285–9.

Shell (2002), *Meeting the energy challenge: The Shell Report 2002*.

World Energy Council (WEC) (1998), 'A keynote address to the 30th Conference of the Japan Atomic Industrial Forum Inc.', by Michael Jefferson, Tokyo, 20 April. *Global Warming and Global Energy after Kyoto*. Available at: http://www.worldenergy.org/wec-geis/publications/default/archives/speeches/spc980420MJ.asp

9 Bootleggers, Baptists and the global warming battle

BBC News (2000), ' "Massive" pollution cuts needed', Nov. 11. Available at: http://news.bbc.co.uk/hi/english/sci/tech/1018874.stm (last visited 12 July 2003).

Black-Arbeláez, T. (2001), *Supplementarity at COP6: Fewer benefits and less development for Asian countries*, paper presented at the International Climate Change Policy session of the National Outlook Conference sponsored by the Australian Bureau of Agricultural and Resource Economics.

Bolin, B. (1998), *The Kyoto Negotiations on Climate Change: A Science Perspective*, 279 *Science*, pp. 330, 331.

Bradsher, K. and Revkin, A. C. (2001), 'Many Companies Cut Gas Emissions to Head Off Tougher Regulations', *New York Times*, 15 May. Available at: http://www.nytimes.com/2001/05/15/business/15ENER.html?pagewanted=print (last visited 12 July 2003).

Breyer, S., *Regulation and its Reform* (Boston, Harvard University Press, 1982).

Brown, Paul (2000), 'Climate talks fail to close rift with US', *Guardian*, 20 Nov. Available at: www.guardianunlimited.co.uk/international/story/0,3604,400163,00.html (last visited 12 July 2003).

Browne, A. (2001), 'Prescott blasts free-rider Bush for Kyoto pull-out', *Observer*, 1 April. Available at: www.observer.co.uk/politics/story/0,6903,466551,00.html (last visited July 27 2001).

Burniaux, J.-M. (1998), *How important is market power in achieving Kyoto?: An assessment based on the GREEN model*, at 1, Paper presented at the OECD workship titled 'Economic Modelling of Climate Change. Available at: www.oecd.org/pdf/M00006000/M00006219.pdf (last visited 12 July 2003).

Cairncross, F., *Green, Inc. : A Guide to Business and the Environment* (Washington, DC, Island Press, 1995).

Castle, S. (2001), 'EU Sends Strong Warning to Bush over Greenhouse Gas Emissions', *Independent*, 19 March, p. 14.

Cooper, R. N. (1998), 'Toward a Real Global Warming Treaty', *Foreign Affairs*, March/April 1998.

Cordato, R. E. (1999), 'Global Warming, Kyoto, and Tradeable Emissions

Permits: The Myth of Efficient Central Planning', Institute for Research on the Economics of Taxation: Studies in Social Cost, Regulation, and the Environment: no. 1 (September).

Cushman, J. H. Jnr. (1998), 'Religious Groups Mount a Campaign to Support Pact on Global Warming', *New York Times*, 15 August, p. A10.

DRI–McGraw-Hill (1997), *The Impact of Carbon Mitigation Strategies on Energy Markets, the National Economy, Industry and Regional Economies*.

Drozdiak, W. (2000), 'Global Warming Treaty Dispute Heats Up; U.S. to Press for Pollution Trading Credits at Hague Meeting, Over European Objections', *Washington Post*, 12 November, p. A26.

The Economist (2001a), 'The next step on global warming', 4 April.

The Economist (2001b), '*Oh no, Kyoto*', 5 April. Available at: www.economist.com/displayStory.cfm?Story_ID=561509 (last visited 12 July 2003).

The Economist (1997), 'Global warming: Rubbing sleep from their eyes,' 13 December, p. 38.

Easterbrook, G. (2001), 'Health Nut', *The New Republic*, 30 April.

Energy Report 27 (2000), 'President wants $4 billion in tax breaks to promote clean energy technologies', 7 February.

Energy Information Administration (EIA), U.S. Dept. of Energy (1998), 'Impacts of the Kyoto Protocol on U.S. Energy Markets and Economic Activity' (1998). Available at: www.eia.doe.gov/oiaf/kyoto/kyotorpt.html (last visited 12 July 2003).

Environment News Service (ENS) (2000), 'Canada's Greenhouse Gas Emissions Prove Tough to Control', 6 September.

European Union (EU), *Green Paper on Greenhouse Gas Emissions Trading within the European Union* (2000). Available at: www.europa.eu.int/comm /environment/docum/0087_en.htm (last visited 12 July 2003).

Exxon-Mobil (2000), Available at: http://www.exxon.mobil.com/Files/ Corporate/000406.pdf

Exxon-Mobil (2001), 'Moving past Kyoto ... to a sounder climate policy,' ExxonMobil (2001). Available at: www.exxon.mobil.com/Files/Corporate/ 170401.pdf and www.exxon.mobil.com/Files/Corporate/170401_1.pdf (last visited 12 July 2003)

Fan, S. *et al.*, 'A Large Terrestrial Carbon Sink in North America Implied by Atmospheric and Oceanic Carbon Dioxide Data and Models', *Science* 282, p. 442 (16 October).

FT Energy Newsletters (2001), International Gas Report, 'Hot air' trading – the Russians are coming?', 16 March.

Geller, Howard. (1998), Executive Director of the American Council for an Energy-Efficient Economy, Testimony before the Science Committee, US House of Representatives (Oct. 9, 1998).

Grubb, M., Vrolijk, C. and Brack, D., *The Kyoto Protocol: A Guide and Assessment* (London, Earthscan, 1999).

Hammitt, J. K. (2000), 'Climate Change Won't Wait for Kyoto', *Washington Post*, 29 November, p. A39.

Horner, C. H. (2000), ' EU Taking Aggressive Stance Toward U.S.', 16 November.

Available at: www.globalwarming.org/hague/cop6horner2.htm (last visited 12 July 2003).

Hourcade, J.-C. and Toman, M., (no date). 'Policies for the Design and Operation of the Clean Development Mechanism'. Available at: www.weathervane.rff.org/research/Toman_policies_CDM.htm (last visited 12 July 2003).

Intergovernmental Panel on Climate Change (1996). *Climate Change 1995: The Science of Climate Change.*

John, M. (1999). 'Is Europe really going non-nuclear', Reuters News Service, 12 July. Available at: www.planetark.org/dailynewsstory.cfm?newsid=2028&newsdate=12-Jul-1999 (last visited 12 July 2003).

Jorgenson, D. W. and Wilcoxen, P. J. (1993), 'Reducing U.S. Carbon Dioxide Emissions: An Assessment of Different Instruments', 115 *J. Pol'y Modeling* 491.

Kakuchi, S. (1997), 'Nuke, Oil Lobbies Bid for Greener Image at Kyoto', Interpress Service, 5 December.

Kyoto Protocol. Available at: www.unfccc.int/resource/docs/convkp/kpeng.html

Laird, F. N. (2000), 'Just Say No to Greenhouse Gas Emissions Targets', *Issues in Science and Technology Online*, National Academy of Sciences (Winter). Available at: www.nap.edu/issues/17.2/laird.htm (last visited 12 July 2003).

Lavelle, M. (2001), *A shift in the wind on global warming*, U.S. News & World Report, 19 March, p. 38.

Leidy, M. P. and Hoekman, B. M., *Pollution Abatement, Interest Groups, and Contingent Trade Policies*, in Congleton, R. D. (ed.), *The Political Economy of Environmental Protection: Analysis and Evidence* (Ann Arbor, University of Michigan Press, 1996).

LeVine, S. (2000), 'Ex-Soviet States Sit on a Gold Mine of Greenhouse Gases', *Wall Street Journal*, 21 November at A23.

Loy, F. (2000), 'Remarks to American Bar Association in London' (July 20). Available at: www.usinfo.state.gov/topical/global/environ/latest/00072103.htm (last visited 12 July 2003).

McKibbin, W. *et al.* (1999), 'Emissions Trading, Capital Flows and the Kyoto Protocol', in Weyant, J. P. (ed.) (1999), 'The Costs of the Kyoto Protocol: A Multimodel Evaluation?', special edition of *The Energy Journal* (International Association for Energy Economics).

Malakoff, M., *Thirty Kyotos Needed to Control Warming*, 278 *Science* 2048 (19 December 1997).

Manne, A. S. and Richels, R. G. (1998), *Economic Impacts of Alternative Emissions Reduction Scenarios*, October. Available at: www.accf.org/manne-richels1098.htm (last visited 12 July 2003).

Manne, A.S. and Richels, R.G. (1999), 'The Kyoto Protocol: A Cost-Effective Strategy for Meeting Environmental Objectives?', in Weyant, J. P. (ed.) (1999), 'The Costs of the Kyoto Protocol: A Multimodel Evaluation?', special edition of *The Energy Journal* (International Association for Energy Economics).

Manne, F. N. and Richels, R. G. (1991), 'Global CO_2 Emission Reductions: The Impacts of Rising Energy Costs', *Energy Journal* 12.

Mitchell, R. B. and Chayes, A. (1995), 'Improving Compliance with the Climate Change Treaty', in Lee, H. (ed.), *Shaping National Responses to Climate Change: A Post-Rio Guide* (Washington DC, Island Press 1995), pp. 115, 120–27

Munday, D. (2000), 'Religious leaders join global warming battle', *The Post and Courier*, 23 Nov. p. B1.

National Energy Information Center (NEIC) (2001), *International Carbon Dioxide Emissions from the Consumption and Flaring of Fossil Fuels (Petroleum, Natural Gas, and Coal) Forecasts*, Appendix A, Table A10 (presenting data from the International Energy Outlook 2001). Available at: www.eia.doe.gov/emeu/international/environm.html (last visited 12 July 2003).

Nelson, R. H., 'Environmental Calvinism: The Judeo-Christian Roots of Eco-Theology', in Meiners, R. E. and Yandle, B. (eds), *Taking the Environment Seriously* (Lanham, MD, Rowman and Littlefield, 1993).

Oil and Gas Journal (2000), 'The collapse of Kyoto', vol. 49, p. 25 (4 December).

Planet Ark (2000), 'Saudi hits out at economic consequences of climate pact', 22 November. Available at: www.planetark.org/dailynewsstory.cfm?newsid=9042 (last visited 12 July 2003).

Raven, G. (1998), 'EU Urges US to Implement Kyoto Global Warming Deal', Reuters News Service (26 April).

Reid, W. V. and Goldemberg, J. (1997), 'Are Developing Countries Already Doing as Much as Industrialized Countries to Slow Climate Change?', *Energy Policy* 26.

'Greenpeace urges climate talks to reject nuclear', Reuters News Service, 6 September 2000. Available at: www.planetark.org/dailynewsstory.cfm?newsid=8059&newsdate=06-Sep-2000 (last visited July 28 2001).

Repetto, R. and Maurer, C. (1997), 'U.S. Competitiveness is Not at Risk in the Climate Negotiations', Climate Notes, World Resources Institute (October). Available at: www.wri.org/wri/cpi/notes/us-comp.html (last visited 12 July 2003).

Reuters News Service (2001a), 'Minister doubts Germany can meet pollution target', 2 April. Available at: www.planetark.org/dailynewsstory.cfm?newsid=10327&newsdate=02-Apr-2001 (last visited 12 July 2003).

Reuters News Service (2001b), 'Climate change may brighten future for nuclear', 2 May. Available at: www.planetark.org/dailynewsstory.cfm?newsid=10679 (last visited 12 July 2003).

Reuters News Service (1999), 'EU's De Palacio says nuclear needed for Kyoto targets,' Reuters News Service, 15 October. Available at: www.planetark.org/dailynewsstory.cfm?newsid=4180&newsdate=15-Oct-1999 (last visited 12 July 2003).

Revkin, A. C. (2001), 'Global Warming's Likely Victims', *New York Times*, 19 February. Available at: www.nytimes.com/2001/02/19/science/19WARM.html (last visited 12 July 2003).

Samuelson, P. A. (1954), 'The Pure Theory of Public Expenditure', *Review of Economic Statistics* 36, p. 387.

Shook, B. (1999), 'Energy Firms Outline Efforts to Pre-empt Kyoto Protocol With

Their Own Initiatives', *Oil Daily*, 7 December.

Steyn, M. (2001), 'Sayonara, Kyoto', *National Post* (Canada), 1 April.

Stott, P. (2001), 'Hot Air + Flawed Science = Dangerous Emissions', *Wall Street Journal*, 2 April, p. A22.

Topping Cone, J. (2000), *EU and US on opposite sides of key issue of Kyoto Protocol*, Earth Times News Service ('The EU, with significant limits on land available for reforestation by its member countries, is lobbying for these credits to have a lower value in an emissions trading system ...').

United States Department of Energy (DOE) (1998), Report # SR/OIAF/98-03, 'Summary of the Kyoto Report'. Available at: www.eia.doe.gov/oiaf/kyoto/kyotobrf.html (last visited 12 July 2003).

US Senate. Res. 98, 105th Cong. (1997).

US White House (2001), 'Text of a Letter from the President to Senators Hagel, Helms, Craig, and Roberts', 13 March. Available at: www.whitehouse.gov/news/releases/2001/03/20010314.html (last visited 12 July 2003).

Utility Environment Report (1997), 'Climate Treaty Negotiators Facing Tough Task in Crafting Amendments', 14 March, p. 9.

Utility Environment Report (1998), 'To Counter Kyoto, West Virginia Prohibits regulation of Greenhouse Gas Emissions', 10 April.

Victor, D. G (2001), 'Piety at Kyoto Didn't Cool the Planet', *New York Times*, 23 March. Available at: www.nytimes.com/2001/03/23/opinion/23VICT.html (last visited 12 July 2003).

Wallsten, S. J. (2000), 'The R&D Boondoggle', Regulation 23, p. 12. Available at: www.cato.org/pubs/regulation/regv23n4/wallsten.pdf (last visited 12 July 2003).

WEFA, Inc. (1998), *Global Warming: The High Cost of the Kyoto Protocol: National and State Impacts: Executive Summary*.

Wigley, T. M. L. (1998), 'The Kyoto Protocol: CO_2, CH_4 and climate implications', *Geophysical Research Letters* 25, p. 2285.

Witter, W. (1997), 'U.S. irks E.U. on pollution controls; Suggests cut based on level of abuse', *Washington Times*, 2 December, p. A9.

Wysham, D., Sohn, J., and Vallette, J. (1999), 'OPIC, Ex-IM, and Climate Change: Business as Usual?', (28 April). Available at: www.foe.org/res/pubs/pdf/climatesummary.pdf (last visited 12 July 2003).

Yandle, B. (1983), 'Bootleggers and Baptists: The Education of a Regulatory Economist', *Regulation* 12 (May/June).

Yandle, B. (1999), *After Kyoto: A Global Scramble for Advantage*, *Independent Review* 4. Available at: www.independent.org/tii/media/pdf/TIR41_Yandle.pdf

10 Climate Change and Civilisation Collapse

Baillie, M., *A Slice Through Time* (London, B. T. Batsford, 1995)

Baillie, M. (1998), Book Review: 'Third Millennium BC Climate Change and Old World Collapse', *Journal of Archaeological Science* 25(2): 185–7

Berglund, B. E. (2003), 'Human impact and climate changes – synchronous events and a causal link?' *Quaternary International* 105, pp. 7–12

Butzer, K. W., 'Sociopolitical Discontinuity in the Near East c. 2200 BCE: Senarios from Palestine and Egypt', in Dalfes, H. N., Kukla, G. and Weiss, H. *Third Millennium B. C. Climate Change and Old World Collapse* (Berlin, New York, Springer, 1996), pp. 245–96.

Claussen, M., Kubatzki, C., Brovkin, V., Ganopolski, A., Hoelzmann, P. and Pachur, H. J. (1999), 'Simulation of an abrupt change in Saharan vegetation in the mid-Holocene, in *Geophysical Research Letters*, vol. 26, no. 14, pp. 2037–40.

Courty, M.-A. (1998), 'The Soil Record of an Exceptional Event at 4000 B. P. in the Middle East', in Peiser, B. J., Palmer, T. and Bailey, M. E. (eds) *Natural Catastrophes during Bronze Age Civilisations*. British Archaeological Reports S728: Oxford, UK.

Cullen, H. M., deMenocal, P. B., Hemming, S., Hemming, G., Brown, F. H., Guilderson, T. and Sirocko, B. (2000), 'Climate change and the collapse of the Akkadian empire: Evidence from the deep sea', *Geology* 28 (4) pp. 379–82.

DeMenocal, P. B. (2001), 'Cultural Responses to Climate Change during the Late Holocene', *Science*, 292 pp. 667–72.

Engvild, K. C. (2003), 'A review of the risks of sudden global cooling and its effects on agriculture', *Agricultural and Forest Meteorology*, 115 pp. 127–37.

Gill, R. B., *The Great Maya Drought. Water, Life and Death* (Albuquerque, University of New Mexico Press, 2000).

Haug, G. H., Gunther, D., Peterson, L. C. D., Sigman, M., Hughen, K. A. and Aeschlimann, B. (2003), 'Climate and the collapse of Mayan civilization', *Science*, 299 pp. 1731–5.

Keys, D., *Catastrophe: An Investigation into to Origins of the Modern World* (London, Century, 1999).

Manning, S. W., 'Cultural Change in the Aegean c. 2200 B. C.', in Dalfes, H. N., Kukla, G. and Weiss, H. *Third Millennium B. C. Climate Change and Old World Collapse* (Berlin, New York, Springer, 1996), pp. 149–72

Mapes, J. (2001), 'Climate Change Linked to Civilization Collapse', *National Geographic News*, 27 February. Available at: http://news.nationalgeographic.com/news/2001/02/0227_climate4.html

Peiser, B. J. (1998), 'Comparative Analysis of Late Holocene Environmental and Social Upheaval', in Peiser, B. J., Palmer, T. and Bailey, M. E. (eds), *Natural Catastrophes during Bronze Age Civilisations*. British Archaeological Reports S728, Archaeopress: Oxford, pp. 117–39.

Pointing, C., *A Green History of the World* (London, Penguin, 1991).

Possehl, G. L. (1996), 'Climate and the Eclipse of the Ancient Cities of the Indus', in Dalfes, H. N., Kukla, G. and Weiss, H., *Third Millennium B. C. Climate Change and Old World Collapse* (Berlin, New York, Springer, 1991), pp. 193–244.

Radford, T. (2001), 'Discovery heralds way for plants to survive drought', *Guardian*, 28 June.

Tainter, J. A., *The Collapse of Complex Societies* (Cambridge, Cambridge University Press, 1988).

Thompson, L. G., Mosley-Thompson, E., Davis, M. E., Henderson, K. A.,

Brecher, H. H. Zagorodnov, V. S., Mashiotta, T. A., Lin, P.-N., Mikhalenko, V. N., Hardy, D. R. and Beer, J. (2002), 'Kilimanjaro Ice Core Records: Evidence of Holocene Climate Change in Tropical Africa', *Science*, 298, pp. 589–93.

Webster, D., *The Fall of the Ancient Maya* (London, Thames & Hudson, 2002).

Weiss, H., Courty, M.-A., Wetterstrom, W., Guichard, F., Seniro, L., Meadow, R. and Curnow, A. (1993), 'The Genesis and Collapse of Third Millennium North Mesopotamian Civilization', *Science*, 261:995-1004.

Weiss, H., 'Beyond the Younger Dryas. Collapse as Adaptation to Abrupt Climate Change in Ancient West Asia and the Eastern Mediterranean', in Bawden, G. and Reycraft, R. (eds), *Confronting Natural Disaster: Engaging the Past to Understand the Future* (Albuquerque, University of New Mexico Press, 2000).

Weiss, H. and Bradley, R. S. (2001), 'What drives Societal Collapse?', *Science*, 291 pp. 609–10.

Wright, K. (1998), 'Empires in the Dust', *Discover Magazine*, March 1998, pp. 22–4.

Yoffee, N. and Cowgill, G. L., *The Collapse of Ancient States and Civilizations* (Tucson and London, The University of Arizona Press, 1988).

11 The political economy of climate change

Baker, R. *Fragile Science. The Reality Behind the Headlines* (London, MacMillan, 2001).

Balling, R. C. Jnr, 'A climate of uncertainty in the greenhouse century', in Morris, J. (ed.), *Sustainable Development. Promoting Progress or Perpetuating Poverty?* (London, Profile Books, 2002) pp. 145–58.

Battaglia, F. (2000), 'Ecco perché l'effetto-serra è solo una grossa bufala,' *Il Giornale*, 4 September.

Bianco, G., Piombini, G. and Stagnaro, C., *Il libro grigio del sindacato* (Bologna, Edizioni Il Fenicottero, 2002).

Bradley, R. L. Jnr, *Julian Simon and the Triumph of Energy Sustainability* (Washington, DC, American Legislative Exchange Council, 2000).

Bradley, R. L. Jnr, 'Energy for sustainable development', in Morris, J. (ed.), *Sustainable Development Sustainable Development. Promoting Progress or Perpetuating Poverty?* (London, Profile Books, 2002), pp. 159–72.

Buchanan, J. M., *Cost and Choice* (Chicago, Marham Press, 1969).

Buchanan, J. M. and Tullock G. (1975), 'Polluters' profits and political response: Direct control versus taxes', *American Economic Review*, 65 pp. 139–47.

Brunetti, M., Maugeri, M. and Nanni, T. (2000),'Variations of Temperature and Precipitation in Italy from 1866 to 1995', *Theoretical and Applied Climatology* 65, pp. 165–74. Available at: www.isac.cnr.it/~climstor/michele/print/TAC65_2000_165.pdf

Castles, I. (2002), 'Letter to Dr. Rajendra Pachauri', August 6. Available at: www.policynetwork.net/pdfs/henderson_castles_letters.pdf.

Cohen, B. L., 'The Hazards of Nuclear Power', in Simon, J. L. (ed.), *The State of Humanity* (Oxford and Cambridge, Blackwell, 1995), pp. 576–87.

Corbyn, P., and Golipur, M., 'What is a Global Temperature? The Over-

---

Representation of Temperate and Polar Zones', in Emsley, J. (ed,), The Global Warming Debate. The Report of the European Science and Environment Forum (Bournemouth, European Science and Environment Forum, 1996).
Cordato, R. (1999), 'Global Warming, Kyoto, and Tradeable Emissions Permits. The Myth of Efficient Central Planning', Studies in Social Costs, Regulation, and the Environment, Institute for Research on the Economics of Taxation, 1. Available at: ftp://ftp.iret.org/pub/SCRE-1.PDF.
Crandall, R. W., 'Economists and the Global Warming Debate', in Adler, J. H. (ed.), The Costs of Kyoto. Climate Change Policy and Its Implications (Washington, DC, Competitive Enterprise Institute, 1997): 145. Available at: http://www.secure.cei.org/PDFs/Costs_of_Kyoto_Part4.pdf
Desrochers, P., 'Industrial ecology and the rediscovery of inter-firm recycling linkages: historical evidence and policy implications', Industrial and Corporate Change 11 (2002): 1031–57.
DRI–WEFA (2002), Kyoto Protocol and Beyond: The High Economic Cost to the United Kingdom. Available at: www.scientific-alliance.com/dri4.doc
Energy Information Administration (EIA) (2002), International Energy Outlook 2002
Office of Integrated Analysis and Forecasting, Washington, DC, Department of Energy. Available at: www.eia.doe.gov/oiaf/ieo/pdf/0484(2002).pdf
European Environment Agency (EEA) (2003), 'EU greenhouse gas emissions rise for second year running.' Available at: www.org.eea.eu.int/documents/newsreleases/ghg-2003-en
European Industrial Relations Observatory Online (EIRO) (no date), 'ETUC anticipates green job creation under Kyoto Protocol', accessed 15 June 2003. Available at: www.eiro.eurofound.ie/1999/04/InBrief/EU9904169N.html
Food and Agricultural Organization of the United Nations (FAO), 1991 Country Tables (Rome: FAO, 1991).
Francescato, G., 'Dal concetto del limite al principio di precauzione', in Francescato, G. and Scanio, A. P., Il principio di precauzione (Milan, Jaca Book, 2002).
Friis-Christensen, E. and Lassen, K. (1991), 'Length of the Solar Cycle: An Indicator of Solar Activity Closely Associated with Climate', Science 254, pp. 698–700.
Gaspari, A. (1997), Profeti di sventura? No grazie! Manifesto per un'ecologia scientifica e ottimista, Milan, 21mo Secolo, 1997.
Gheddo, P. and Beretta, R., Davide e Golia. I cattolici e la sfida della globalizzazione (Milan, San Paolo, 2001).
Goklany, I. M. (2001a), Economic Growth and the State of Humanity (Bozeman: Political Economy Reseach Center, 2001).
Goklany, I. M. (2001b), The Precautionary Principle (Washington, DC, Cato Institute, 2001).
Goklany, I. M. (1995), 'Strategies to Enhance Adaptability: Technological Change, Sustainable Growth and Free Trade', Climatic Change 30.
Hardin, G. J. (1968), 'The Tragedy of the Commons', Science 162, pp. 1244–5.
Infantino, L., Individualism in Modern Thought. From Adam Smith to Hayek (London, Routledge, 1998).

Intergovernmental Panel on Climate Change, *Climate Change 2001: The Scientific Basis* (Cambridge, Cambridge University Press, 2001).

International Energy Agency (IEA), *Energy Policies of IEA Countries* (Paris, OECD/IEA, 2001).

International Monetary Fund (IMF), *World Economic Outlook: April 2000* (Washington, DC, IMF, 2000).

Jones, P. D., Parker, D. E., Osborn, T. J. and Briffa, K. R. (2001), 'Global and hemispheric temperature anomalies – land and marine instrumental records', in *Trends: A Compendium of Data on Global Change*. Carbon Dioxide Information Analysis Center, Oak Ridge National Laboratory, U.S. Department of Energy, Oak Ridge, Tenn., USA. Available at: http://www.cdiac.esd.ornl.gov/trends/temp/jonescru/jones.html.

Kyoto Protocol to the United Nations Framework Convention on Climate Change, 1997.

Leoni, B., *Freedom and the Law* (Indianapolis, Liberty Fund, 1991).

Lomborg, B., *The Skeptical Environmentalist. Measuring the Real State of the World* (Cambridge, UK, Cambridge University Press, 2001).

McKitrick, R. (2001), 'The Influence of Economic Activity on the Measurement of Global Warming', September. Available at: www.uoguelph.ca/~rmckitri/research/gdptemp.pdf

Marsh, G. E. (2002), 'A Global Warming Primer', The National Center for Public Policy Research, *National Policy Analysis* 420. Available at www.nationalcenter.org/NPA420.pdf

Meiners, R. E. and Yandle, B. (1998), 'Common law environmentalism', *Public Choice* 94 pp. 63–64.

Midena, M. (2002), 'Il lavoro sarà sempre più verde, parola di Sindacato', *Ambiente e lavoro*, July–August. Available at www.cisl.it/universita/2002/Rivista/7_02/lavoro.pdf

Mitchell, W. C. and Simmons, R. T., *Beyond Politics* (Oakland, CA, The Independent Institute, 1994), pp. 46–162

Montgomery, W. D. (1997), 'Global Impacts of a Global Climate Change Treaty', in Adler, J. H. (ed.), *The Costs of Kyoto*, pp. 65–8. Available at www.secure.cei.org/PDFs/Costs_of_Kyoto_Part2.pdf

Moonen, A. C. *et al.* (2002), 'Climate change in Italy indicated by agrometeorological indices over 122 years', *Agricultural and Forest Meteorology* 111, pp. 13–27.

Nordhaus, W. and Boyer, J. (1999), 'Requiem for Kyoto: An economic analysis of the Kyoto Protocol', *The Energy Journal: Kyoto* special issue, pp. 93–130. Available at: http://www.econ.yale.edu/~nordhaus/homepage/Kyoto.pdf

Ortolani, F. (2002). 'Modificazioni climatico ambientali cicliche tipo "effetto serra" durante il periodo storico', *21mo Secolo* 3.

Parry, M., Rosenzweig, C., Iglesias, A., Fischer, G. and Livermore, M. (1998), 'Buenos Aires and Kyoto targets do little to reduce climate change impacts', *Global Environmental Change* 8, vol. 4, pp. 285–89.

Reisman, G. (2002), 'Environmentalism in the Light of Menger and Mises', *The Quaterly Journal of Austrian Economics* 5, pp. 13–14.

Ricci, R. A. (2002), 'Problemi ambientali ed informazione scientifica', *Nuova*

Secondaria 10.
Rothbard, M. N., *For a New Liberty. The Libertarian Manifesto* (Auburn, Ala., The Ludwig von Mises Institute, (2002), pp. 260–8. Available at: http://www.mises.org/rothbard/newliberty.asp
Serafini, M. (1998), 'Relazione introduttiva al Consiglio Nazionale di Legambiente', November.
Simon, J. L. (1996), *The Ultimate Resource* 2 (Princeton, Princeton University Press, 1996).
Soon, W. and Baliunas, S. *et al.*, *Global Warming. A Guide to Science* (Vancouver, The Fraser Institute, 2001).
Soon, W. *et. al.* (2003), 'Reconstructing climatic and environmental changes of the past 1000 years: A reappraisal', *Energy and the Environment*, vol. 14, nos. 2&3, pp. 233–96.
Spencer, R. W. and Christy, J. R. (1990), 'Precise Monitoring of Global Temperature Trends From Satellite', *Science* 247, p. 1558.
Steele, C. N., 'The Soviet Experiment: Lessons for Development', in Morris, J. (ed.), *Sustainable Development. Promoting Progress or Perpetuating Poverty?* (London, Profile Books, 2002) pp. 88–103.
Thorning, Margo (2002), *Kyoto Protocol and Beyond: Economic Impacts on EU Countries*, Washington, DC, American Council for Capital Formation. Available at: http://www.accf.org/ACCF_KyotoEconImp.pdf
Thorning, Margo (1998), 'Climate Mitigation Policy and US Economic Growth'. Testimony before the Congressional Subcommittee on National Economic Growth, Natural Resources, and Regulatory Affairs, US House Committee on Government Reform and Oversight, 23 April 1998. Available at http://www.accf.org/Apr98test.htm
Tognetti, R. *et al.* (1998), 'Transpiration and stomatal behaviour of *Quercus ilex* plants during the summer in a Mediterranean carbon dioxide spring', *Plant, Cell and Environment* 21, pp. 613–22
World Energy Council (WEC) (1998), 'A keynote address to the 30th Conference of the Japan Atomic Industrial Forum Inc.', by Michael Jefferson, Tokyo, 20 April. *Global Warming and Global Energy after Kyoto*. Available at: http://www.worldenergy.org/wec-geis/publications/default/archives/speeches/spc980420MJ.asp
Yandle, B. (1999), 'After Kyoto: A Global Scramble for Advantage'. *Independent Review* 1; pp. 19–40. Available at: http://www.independent.org/tii/media/pdf/TIR41_Yandle.pdf

Epilogue
Daily News (Sri Lanka) (2003) ,'Preparing for a future deluge', 2 July. Available at: http://www.dailynews.lk/2003/07/02/fea01.html
De Soto, H., *The Other Path: The Invisible Revolution in the Third World* (New York, Harper and Row, 1989).
Dresner, S., *The principles of sustainability* (London, Earthscan, 2002) p. 58.
European Commission (2002), 'SMEs in Focus: 2002 Observatory of European

SMEs'. Available at: http://europa.eu.int/comm/enterprise/enterprise_policy/
analysis/doc/execsum_2002_en.pdf

Friends of the Earth UK (FOE UK) (2001), 'More from Less'.

IFG (2002). *Alternatives to Economic Globalization*, Spring.

Louw, L. (2002), 'Poverty today is truly miraculous', *Sunday Telegraph*, 1
September. Available at: http://www.telegraph.co.uk/
opinion/main.jhtml?xml=/opinion/2002/09/01/do0101.xml

Pearce, F., *Green Warriors: The People and Politics Behind the Environmental
Revolution* (London, Bodley Head, 1991), p. 284.

Retallack, S, and Sobhani, L. (2002), 'Atmosphere: Climate and Ozone', *Intrinsic
Consequences of Economic Globalization on the Environment*, San Francisco,
International Forum on Globalization, August.

Acknowledgements

My utmost gratitude goes to the authors for their excellent contributions and for following a demanding schedule. One contributor, Julian Morris, deserves special mention for helping to conceptualise the book and for providing many helpful changes and corrections. Many thanks to Linda Whetstone, the Chairman of International Policy Network, who provided many helpful edits in the eleventh hour. Similarly, I thank the trustees of International Policy Network for their flexibility and enthusiasm for all of IPN's work. Robert Dimery was a speedy and reliable copy-editor. As ever it was a pleasure to collaborate with Paul Forty and the staff of Profile Books. Finally, I give thanks and love to my families on either side of the Atlantic who constantly extend an admirable amount of patience, and inspiration, to ideologues.

Kendra Okonski
London
September 2003

Index

Page numbers in *italic* refer to Tables and Figures

aluminium 138
Ammerman, Albert 47–8, 49,
  50
Ancient Greece, malaria in 24,
  25
Ancient Rome, malaria in 24–5
Anderson, David 174
Anopheles atroparvus 28, 33–4
Anopheles maculipennis 33–4
*Anopheles maculipennis sensu
  strictu* 33–4
Anopheles messeae 33–4
Antarctica 61
apartheid 116, 117, 121, 128
archaeological evidence 196
Australia 134, 184, 220

Baillie, Michael 196
Baker, Robin 204, 207–8
Bangladesh 227–8
'Baptists' 177–8, 185
Basel Convention (1989) 140–1
Battaglia, Franco 206
Belgium 183
benefit-sharing 16, 230–1
biodiversity 56, 71, 72
  loss 11, 20, 105, 106, 107,
    223, 226; threats to 59, 61
biomass energy 11, 98, 99,
  100–2, 114, 122–3, 226
  efficiency 101; environmental
    costs 105–7; health impacts
    101, 102–4, *102*, *104*; use in
    India 100, *100*, 101, *101*
biotechnology 200, 200–1
birds, impact of warming on
  60
Body Shop, The 98
'bootleggers' 178, 190

'bootleggers and Baptists'
  theory of regulation 14,
  176–8
BP/Amoco 158, 160–1, 161,
  163
Breyer, Stephen 178–9
Bronze Age cultures 196
'bubble' concept (EU) 175, 178,
  187–8
Buck, Stuart 228
bureaucracy 69, 120, 121, 128,
  136, 230, 231
Bush, George W. 173, 185
business 13, 114, 162, 165,
  167, 228
  as agent of change 13, 169,
    170; bootlegging role 181–2;
    effects of regulation on 13,
    157–60, 164, 168; efficiency
    13, 165, 229; and
    environment 164–6; and EU
    152–3, 169; 'greenwashing'
    13, 162–4; influence on
    regulations 13, 161, 167,
    170; relocation 159, 170,
    181; response to climate
    change 13, 160–2; *see also*
    large businesses; SMEs
Butzer, Karl W. 196, 197
by-product recovery 164–5
Byzantine Empire 195

Calder, Nigel 118
Canada 184
carbon dioxide *see* CO_2
carbon sinks 65, 71, 92, 175,
  175–6, 186
carbon stores 65, 71
carbon taxes 182–3

Global Climate and Energy
  Project *see* G-CEP
global cooling 118–19, 169,
  193, 200
Global Environment Facility *see*
  GEF
global industrial policy 189–90
global population 57, 199–200
global warming 5–6, 8–10, 21,
  190, 204, 221, 223
  acceptance of 16, 133, 154,
    220; adaptation approach
    16, 211, 219; benefits 70,
    207; and civilisation collapse
    15–16, 192, 199–200;
    'climate control' approach
    16, 218–19, 221, 221–2,
    227–8, 231; costs 209;
    dealing with 12–14, 74,
    95–7, 216–17; debate over
    54–5, 203, 219–23; effect on
    agriculture 88; effect on
    health 88–9; effect on human
    welfare 86–9; estimated
    future impacts 56, 61–5, 63;
    exaggeration of risks 19–20,
    202; impacts up to the
    present 57–61; IPCC
    forecasts 79, 84–6, 85; as
    last straw 65–9, 72–4; as
    necessary myth 178;
    suggested responses to 74,
    216–17; timescale 208–9,
    209; *see also* climate change;
    climate-sensitive indicators;
    Kyoto Protocol; sea level
    rises
globalisation 98, 115, 159,
  221

Goklany, Indur 8, 224, 225,
  229–30
gonotrophic dissociation 34
governments 167
  accountability 69, 231; and
    energy provision 115;
    intervention by 214, 215; *see
    also* subsidies
Greece 88, 188
  malaria in 33
Greece (Ancient), malaria in 24,
  25
Green Party (Italy) 39, 203
  opposition to Project MOSE
    45–6, 47, 54
greenhouse effect 6, 22, 204,
  219–20
greenhouse gases *see* GHGs
Greenland 205
Greenpeace 41, 98, 162–3
'greenwashing' 13, 162–4
grid electricity 125–6, 128
Gross Domestic Product *see*
  GDP
growing season 60, 88

habitat tracking 200
  see also relocation
habitats
  loss 59, 61, 65, 71, 71, 73,
    226; for mosquitoes 24, 28,
    30, 33, 34
Hansen, James 219–20
Harvey, William (1578–1657)
  27
hazardous waste 140–1
health
  and climate 20; effect of global
    warming 88–9; *see also*

criticisms of 83–4, 85–6, 88,
207–8; economic forecasts
208; emissions scenarios
82–4, *85*, 86; estimate of
costs of warming 86–7, 95;
global warming forecasts 79,
84–6, *85*, 206; malaria
forecast 36; prediction
models 79, 80–1; sea level
rise predictions 48, 57–8;
Third Assessment Report
(TAR) 2001 61–2, *63*, 79
Italia Nostra 39, 43, 46, 47, 53
Italy 93, 212, 213
  effect of Kyoto Protocol 203;
  emissions 203; global
  warming debate 202–3;
  malaria in 23, 32, 33, 35;
  warming 203; *see also* Venice

Japan 107–8, 212
Joint Implementation 175–6
Jorgensen, Dale W. 180

Kazakhstan 187
Kenny, Andrew 12
kerosene *see* paraffin
Kyoto Protocol (1997) 6, 77–8,
  89, 134, 155, 174–5, 211–12
  benefits 10, 69–70, 89–90, 95;
  beyond 2012 94–5;
  'bootleggers and Baptists'
  negotiations 14, 176–8; cost-
  efficiency of 97; costs 10,
  69–70, 77, 90, 90–5, 95–6,
  202, 211–14, 216; and
  economic growth 155, 213;
  effects on different countries
  158–9, 180, 181, 212–13;

effects on different industries
180–1; enforcement 134,
135–7, 147, 189; estimated
impacts 62–5, *63*, 95, 155,
176; EU and 15, 154, 211,
213–14; first commitment
period (2008–12) 15, 91–4,
154, 174–5; implications of
Shrimp–Turtle dispute
146–7; Joint Implementation
175–6; justification 215;
limited effect 174, 176,
211–12, 216, 221; and
malaria 72; media discussion
of 203; opposition from
industry 184; redistribution
effects 185–6; super-Kyoto
regime 212; supporters
176–7, 183; trade
restrictions against non-
ratifying parties 13, 14, 148,
149, 151; USA failure to
ratify 77, 92, 134, 155,
173–4, 176; USA
requirements under 174–5;
see also CDM; CO$_2$; global
warming

land conversion 59, 65, 71, 71,
226
large businesses 13, 176
last straw, climate change as
65–9, 72–4
less developed countries *see*
poor countries
life expectancy 57, 68, 68, 73,
128
Lindzen, Richard 81
liquid petroleum gas *see* LPG